NOW AVAILABLE

The new, completely revised 1995 edition of

INTERNATIONAL CODE OF NOMENCLATURE FOR CULTIVATED PLANTS

(ICNCP OR CULTIVATED PLANT CODE)

P. Trehane, C.D. Brickell, B.R. Baum, W.L.A. Hetterscheid, A.C. Leslie, J. McNeill, S.A. Spongberg, & F. Vrugtman (ed.), *The International Code of Nomenclature for Cultivated Plants – 1995* (ICNCP or Cultivated Plant Code), adopted by the International Commission for the Nomenclature of Cultivated Plants. Quarterjack Publishing, Wimborne, U.K., 1995 (ISBN: 0 948117 01 X).

....

The international set of rules for naming agricultural, forestry and horticultural plants is now fully revised in a thoroughly updated 6th. edition for use by everyone concerned with the accurate naming of plants.

The *Code* is prepared and released under the authority of the International Commission for the Nomenclature of Cultivated Plants.

The last edition was issued in 1980 and has long been out of print. Since the preparation of the 1980 *Code*, there have been many changes within the applied disciplines of agriculture, forestry and horticulture. In particular the introduction of national and international legislation and its accompanying statutory requirements have had great impact in many countries as have considerations of intellectual property rights, the use of information technology and the steady expansion of the range of International Registration Authorities.

The design of the *Code* is dramatically different from previous editions. From being a short pamphlet, the *Code* has grown into a 192 page book containing a number of extensive appendices in addition to the Rules to make this a veritable *vade mecum* for all those needing to form and maintain the correct names for man's own plants.

The *Code* has been designed with the particular needs of plantsmen, the nursery trade, raisers of new plants, compilers of plant directories, authors, journalists, editors, teachers, taxonomists, governmental bodies and registration authorities in mind. It is not an exaggeration to claim that this is a truly indispensable manual for all who take their profession or hobby seriously.

The Rules of the *Code* are based on a set of 12 governing Principles and are divided into 32 Articles which deal with the formation, maintenance and use of the names of three taxonomic groups of cultivated plants; the cultivar, the cultivar-group and the graft-chimaera.

The provision of a number of appendices focusing on the work of registration authorities, herbaria maintaining Standards, a "nomenclatural filter" for deciding if a name is correct or not, a quick guide for forming new cultivar names, a résumé of how Latin names of plants are formed, an extensive list of checklists of cultivar names and a comprehensive glossary of terms used in nomenclature, along with indices to scientific names and subject, complete the thoroughness behind this edition.

This edition of the *Code* is designed to complement, yet be independent from, the *International Code of Botanical Nomenclature* and should remain operative into the next century.

ORDERING DETAILS

The *Code* is available for £18.50 + £1.50 post and packing within the UK (£20.00)

For world-wide surface mail and air-mail to Europe, post and packing is £2.00 (£20.50)
For air mail outside Europe, post and packing is £3.50 (£22.00)
For air mail to New Zealand, Australia, Japan and the Far East, post and packing is £4.00 (£22.50)

Send your name and address, with completed cheque or money order made payable to "Quarterjack Publishing" (£UK only) to:

Quarterjack Publishing, Hampreston Manor, Wimborne, Dorset BH21 7LX, United Kingdom

INTERNATIONAL CODE OF NOMENCLATURE
FOR CULTIVATED PLANTS – 1995

Regnum Vegetabile

a series of publications for the use of plant taxonomists published under
the auspices of the International Association for Plant Taxonomy

edited by Werner Greuter

Volume 133

ISSN 0080–0694

International Code of Nomenclature for Cultivated Plants – 1995

(ICNCP or Cultivated Plant Code)

Adopted by the International Commission for
the Nomenclature of Cultivated Plants

Prepared and edited by

P. TREHANE

and

C. D. BRICKELL (Chairman), B. R. BAUM,
W. L. A. HETTERSCHEID, A. C. LESLIE,
J. McNEILL, S. A. SPONGBERG, F. VRUGTMAN,
Members of the Editorial Committee

1995

QUARTERJACK **QP** PUBLISHING

WIMBORNE, UK

ISBN 0 948117 01 X

Printed and bound in Great Britain by Bookcraft (Bath) Ltd.

Cover illustration by Geoff Stone

CONTENTS

FOREWORD

This is the 6th. edition of the International Code of Nomenclature for Cultivated Plants. The first edition in 1953 was succeeded by further editions in 1958, 1961, 1969, and 1980. The history of nomenclature of cultivated plants and of the early *Codes* has been provided by Professor W. T. Stearn in his introduction to the 1953 *Code* and more recently in *Acta Horticulturae* 182. 18-28. 1986.

The 42 years since the *Code* was first published have seen many changes in which the Rules of the 1953 *Code* have been adapted and developed to meet the needs of users of the names of cultivated plants. Many factors during this period have affected the ways in which names are applied to cultivated plants and the introduction of international and national legislation and its accompanying statutory requirements have had a great impact in many countries as have considerations of intellectual property rights, the use of information technology and the steady expansion of the range of International Registration Authorities. Although the Principles of the 1953 and subsequent *Codes* have been fully maintained and strengthened continuously during this period, the formation of two groups of horticultural taxonomists in Europe during the last seven years has given great impetus to the long-recognized need to develop further the pioneering work of Stearn and his colleagues to meet the rapid changes in the breeding, propagation and uses of cultivated plants and the application of uniform and stable names to them.

In March 1988 a number of horticultural taxonomists based at RBG Kew, RBG Edinburgh, Glasnevin and the Royal Horticultural Society, formed the Horticultural Taxonomy Group (HORTAX). A few others with an interest in horticultural taxonomy were later invited to join the group and this informal association set themselves the objective of drawing up a proposal for a revision of the ICNCP. A early draft of their proposal was circulated around the world for comment prior to the ISHS International Horticultural Congress in Florence in 1990 and this created much interest and stimulating responses from many people working with cultivated plants.

Later in 1988 a similar group in the Netherlands also began working on a proposal. The Nomenclature and Registration Working Group of the Vaste Keurings Commissie (VKC) issued an early draft of their proposal in mid-1990.

On 14th December, 1990 an historic meeting took place at the Royal Botanic Gardens, Kew between the members of HORTAX and VKC which, although identifying problem areas in the two independent approaches towards a new *Code*, also realised much common ground.

HORTAX presented its final proposal to the Commission in April 1992 and VKC presented its independent proposal in February, 1994. Although there was a different philosophical approach to each proposal, both had very much in common thanks to the Kew meeting and to the free exchange of ideas that followed that meeting.

Piers Trehane, one of the members of the Commission, was asked to combine the two proposals along with the existing provisions of the *Code* to provide a working document for the next meeting of the Commission. This synthesis, and the overall design for a new *Code*, was unveiled at the 2nd ISHS Symposium on the Taxonomy of Cultivated Plants held at the University of Washington, Seattle in August 1994.

Following the symposium, the Commission met for two and a half days from August 17th 1994 during which time the Principles to underpin the revised *Code* were formulated and much of the detail refined.

The members of the Commission present at Seattle formed the Editorial Committee for the *Code* which was placed under the general editorship of Piers Trehane.

Apart from producing and defining a set of Rules for the naming of cultivated plants, the new *Code* is a publication that provides a great deal of information on other aspects of cultivated plant nomenclature, most of which has been difficult to locate or unavailable previously.

The introduction of a number of appendices covering International Registration Authorities, statutory plant registration authorities, denomination classes, a list of conserved epithets, herbaria maintaining Standards and checklists of ornamental cultivars, as well as a substantial glossary, should provide a *vade mecum* for all interested in cultivated plant nomenclature. A particularly important feature is the inclusion of a system of Standards akin to, but not identical with, the type method required under the rules of the ICBN, which it is hoped will stimulate further the Principles set out in this *Code*. This system was discussed initially at the 1st ISHS Symposium on the Taxonomy of Cultivated Plants held at Wageningen in 1985 and has been formulated for the first time in this *Code*.

This edition marks the start of a new era in the 42 year history of this *Code*. Any code which has undergone such a major change in design will have its short-comings and these will only be discovered when the Rules of the *Code* are put to the test. Changes and refinements will have to be made in the light of experience in all disciplines. The Commission will meet following the next ISHS International Symposium on Cultivated Plant Taxonomy to be held in the UK in 1998 prior to the XXVth International Horticultural Congress in Brussels in that same year.

Proposals to modify this *Code* should be sent to Piers Trehane, Hampreston Manor, Wimborne, Dorset, BH21 7LX, UK who will act as *rapporteur* to the Commission on matters relating to the *Code*. All other business of the Commission, including proposals to conserve cultivar epithets, should be addressed to the Secretary of the Commission, Wilbert Hetterscheid, Vaste Keurings Commissie, Linnaeuslaan 2a, 1431 JV Aalsmeer, The Netherlands.

I would like, on behalf of the Commission, to record my very warmest thanks to Piers Trehane who has laboured long and hard and given very freely of his time to the overall editing and preparation of this edition of the *Code*. His dedication and patience has been greatly appreciated by all members of the Editorial Committee.

The Commission has also been most fortunate in having Professor John McNeill join them as the representative of the International Union of Biological Sciences (IUBS). His experience with codes of nomenclature generally has been invaluable and his generosity with his time at the Seattle meeting and subsequently, during the editorial process, is very gratefully acknowledged.

Finally I would like to thank all those members of HORTAX and VKC who have made such a contribution to the progress of this *Code*. A special note of thanks is directed to Willem Brandenburg of the Netherlands who was not able to attend the meetings in Seattle, but who has dedicated so much time and effort to matters of international plant nomenclature.

C. D. Brickell
Chairman of the Commission and of the Editorial Committee.

IUBS INTERNATIONAL COMMISSION FOR THE NOMENCLATURE OF CULTIVATED PLANTS

Members at 14th August 1994

representing the interests of
Agriculture (A), Forestry (F), Horticulture (H)

Chairman
Mr C. D. Brickell
The Camber
The Street
Nutbourne
Pulborough
West Sussex RH20 2HE
United Kingdom

Professor J G Hawkes (A)
Dept of Plant Biology
University of Birmingham
P O Box 363
Birmingham B15 2TT
United Kingdom

Secretary
Drs W. L. A. Hetterscheid
Vaste Keurings Commissie
Linnaeuslaan 2a
NL–1431 JV Aalsmeer
The Netherlands

Professor He, Shan-An (H)
Nanjing Botanical Garden
Mem Sun Yat-Sen
Nanjing
Jiangsu
People's Republic of China

Dr B. R. Baum (A)
Land & Biological Resources Research
Neatby Building B33
Central Experimental Farm
Ottawa
Ontario
Canada K1A 0C6

Dr R. D. Johnston (F)
CSIRO
Division of Forest Research
Canberra
Australia

Ir W. A. Brandenburg (H)
CPRO–DLO
Postbus 16
NL–6700 AA Wageningen
The Netherlands

Professor P. Lemattre (H)
Chaire de Cultures Florales
ESHS
4 Rue Hardy
F–78000 Versailles
France

Dr W. B. Critchfield (F)
Pacific Southwest Forest & Range
  Experimental Station
PO Box 245
Berkeley
California 95701
United States of America

Dr A. C. Leslie (H)
The Royal Horticultural Society's Garden
Wisley
Woking
Surrey GU23 6QB
United Kingdom

Members of the Commission

PREFACE

This 1995 edition of the International Code of Nomenclature for Cultivated Plants is strikingly different in appearance from previous editions. Many taxonomists dealing with cultivated plants are used to working within the successful format of the International Code of Botanical Nomenclature (*ICBN*) and this *Code* follows that model.

There is an interlocking relationship between the two *Codes* in so far as they both deal with the scientific names for plants. The *ICBN* essentially deals with the formation and use of plant names in Latin form, whereas this *Code* (except for graft–chimaeras) deals with the final part of the name, the epithet, used for what might be termed man–made plants and which, since 1958 has to be formed in a modern language.

Codes of nomenclature have to blend purposes of precision with an ease to read. This *Code* has to use a certain terminology for reasons of precision and readers should not be intimidated by the constant and persistent use of seemingly foreign terms; they are used for exactness and it is hoped that those concerned with the naming of cultivated plants, whether they are professional or amateur plantsmen, will come to understand this requirement. For those uncertain about the meaning of a word or phrase, the extensive glossary provided as the final Appendix should be helpful.

This *Code* was designed with a number of user–groups in mind. The plant taxonomist, statutory or non–statutory registration authorities, the seed and plant trade (in the widest possible sense), those who compile accurate lists of names for whatever purpose, teachers of plant sciences, journalists, librarians and historians are just some of the major groups who work in, or on the fringes of, agriculture, forestry and horticulture and who require the international standards for nomenclature that this *Code* provides.

This *Code*, as all other codes, has to deal with sorting out problems with the nomenclature of the past as well as providing for that of the future and many of the rules are designed to solve such problems. Users of this *Code* may feel irritated by the need to include what appear to be arcane rules, but they are a real necessity for those having to trawl the historical record to provide accurate and stable names. Those wanting a simple guideline on the creation of new cultivar names are referred to Appendix VIII, but they are advised that this quick guide should be used in conjunction with the main body of rules and not as a replacement for it.

This *Code* deals with the definition, formation and use of three taxonomic groups of cultivated plants; the cultivar, the cultivar–group and the graft–chimaera. The order of the nomenclatural process for fixing the names of these groups is (1) publication, (2) establishment and (3) acceptance.

Publication, establishment and acceptance are the criteria laid out in all codes of nomenclature although, historically, different codes have used a variety of terminology to embrace these concepts.

In March 1994, one of a series of meetings under the auspices of the International Union of Biological Sciences (IUBS) between representatives of the various codes of biological nomenclature took place in Egham, Surrey, UK with a view to drafting a proposal for a unified Code of Bionomenclature which, it is proposed, will eventually provide for future names of all organisms. One of the results of that meeting was to recommend that the various codes standardize on the terminology for the various nomenclatural acts and processes and this edition of this *Code* adopts this standard terminology.

Publication (previously "effective publication") is the first act in fixing a name or epithet. Unless a name or epithet has been properly published, it is not considered to exist in the first place.

Establishment (previously "valid publication") provides for the publication being dated, the name or epithet being in a certain form, and the insistence that a new name or epithet is accompanied by a description.

Only when the various conditions of publication and establishment are fully met, can one decide on the **acceptance** of a name or epithet (previously whether or not it is "correct").

In addition, the term **precedence** is used in place of "priority".

There are a number of other important changes to previous editions of this *Code*.

- At the outset, this *Code* sets down the limits of its intended authority by defining what is meant by "cultivated plants". (Principle 2)

- The principle is stated that new names for taxonomic groups of cultivated plants are to be formed under the provisions of this *Code* (Principle 2) and the recommendation is made that such taxonomic groups below the rank of genus should no longer be named under the provisions of the *ICBN*. (Rec. 16A)

- The relationship between this *Code* and the *ICBN* is more clearly stated. (Principle 2, Art. 1.1, 16.4 etc.)

- The Articles previously covering the formation of Latin names are consigned to the new Appendix IX.

- The rules for the formation of cultivar epithets have been widened considerably with many of the previous restrictive rules and recommendations being replaced by rules designed to avoid the creation of confusing, or potentially confusing, epithets. In particular, for new epithets, there in no longer a limitation on the number of words in an epithet. (Art. 17.10)

- The "cultivar-group" (previously "group") is more closely defined and a number of rules provide for the formation of cultivar-group epithets. (Art. 4 & 19)

- The "grex" becomes a proscribed term, except for its use in orchid nomenclature. (Art. 4.6)

- The previously-used term "cultivar class" has been changed to "denomination class" to parallel the latter use in legislative texts. (Art. 6)

- The starting point for nomenclature has been altered. Previous editions of this *Code* had stated that it was to be the very scarce Philip Miller's *Gardeners Dictionary*, Ed. 6, 1752. The starting date now runs parallel to that specified in the *ICBN*, Linnaeus's *Species Plantarum* of 1753. (Art. 13.1)

- The duties and powers of International Registration Authorities have been specified (Art. 7, 14.2-14.5). An extensive guide for IRAs has been provided. (Appendix I)

- The relationship between cultivar epithets and "trade designations" (previously "commercial synonyms") which are sometimes used instead, is exemplified. (Art. 11)

– Translations of cultivar epithets are no longer permitted; if they occur they are to be termed trade designations. (Art. 28.1)

– International standards for transliteration have been laid down. (Art. 28.4–28.5)

– Provision has been made to conserve epithets that would normally not be accepted yet which *de facto* are in common use and whose replacement by acceptable epithets would cause confusion and do little to enhance stability in nomenclature (Art. 14.1). Appendix V, the list of conserved epithets has been opened.

– Very stringent conditions are laid down to provide for re-use of a cultivar epithet. (Art. 26)

– The concept of "Standards" is formally introduced for the first time. (Art. 12)

It is recognized that this *Code* could be improved in a number of areas and such adjustments are likely to centre around clarification of the following issues:

Some may feel uncomfortable with the term "taxon" when used as a concept in the nomenclature of cultivated plants. The new term "culton" was casually coined by the Chairman of the Commission for this *Code* at an ISHS meeting in Davis, California in 1986 to differentiate between taxa of wild plants and those of cultivated plants and a philosophical rationale for the culton concept has been published by workers in the Netherlands. Whole-hearted and international adoption of this term has yet to be demonstrated but the word "culton" may prove to be a useful formal abbreviation for the cumbersome phrase "taxonomic group of cultivated plants as defined in this *Code*".

What consitutes a "cultivated plant" needs further examination and the range of categories to be admitted as cultivars might be widened.

More work will be required on the application and use of the cultivar-group. Initial rules for the formation of cultivar-group epithets are provided for in this *Code*, but working experience may demonstrate the need for more stringent requirements.

Attention must be given to the rules governing the demarcation of cultivar epithets. This *Code* permits only one method of designation, enclosure with single quotation marks (although a permitted minor typographical variation is allowed). In the ever-changing world of general editorial practice, single quotation marks are being used more and more in place of the more traditional double quotation marks and the demarcation of cultivar epithets becomes harder to recognize. In addition, the inclusion of apostrophes in an epithet may appear visually confusing to some. Should the *Code* permit an alternative device for marking out cultivar epithets within a name?

Although this *Code* provides standards for transliteration from scripts using a non-Latin alphabet into those using Latin, much more should be done to provide standards to operate in reverse if this *Code* is to be perceived as being truly international.

The operation of a system of Standards is in its infancy and difficulties may well come to light which may need to be ruled on in later editions of this *Code*.

These and other outstanding areas, will be developed more fully in further editions of this *Code*.

Piers Trehane
for the Editorial Committee

COMPARISON BETWEEN 1980 *CODE* AND THE PRESENT *CODE*

This key demonstrates what happened to the Articles, Notes and Recommendations in the 1980 *Code*.

1980 Code	1995 Code	1980 Code	1995 Code
GENERAL CONSIDERATIONS & GUIDING PRINCIPLES		17	Appendix IX
		18	deleted
		18A	deleted
1	Prin. 1	18B	deleted
2	Prin. 2	18 Note 1	deleted
3	Prins 1, 6, 9	19	25.2 p.p.
4	Prins 9, 10	19A	deleted
5	Prin. 11		
6	Prin. 12	**Graft–chimaeras**	
CATEGORIES & THEIR DESIGNATIONS		20	5.1, 20.1, 20.3
		21	20.2
		22	20.5
Main Categories		23	20.3
		24	20.6
7	1.1		
8	Appendix IX	**Supplementary categories**	
9	Appendix IX		
10	2.1, 2.2	25	Appendix IX
10 Note 1	2.17	26	4.1, 4.2
10 Note 2	2 Note 1	26 Note 1	deleted
10 Note 3	2 Note 2 p.p.		
10 Note 4	2 Note 4	**FORMATION OF CULTIVAR NAMES**	
10 Note 5	16.2, 16.3		
10 Note 6	2.15		
11	2.5		
11a	2.6, 2.10	27a	17.1, 17.9
11b	1.12, 2.13	27b	17.3
11c	2.11	27c	14.3
11d	2.14	28	29.1, 30.1
11e	2.7, 2.8, 2.9	29	17.6, 17.7, 31.1
12	2.4	30	deleted
12A	deleted	31a	17.14
		31b	deleted
Collective names		31c	17.16
		31Aa	deleted
13	Appendix IX	31Ab	deleted
14	Appendix IX	31Ac	deleted
14A	Appendix IX	31Ad	deleted
15	Appendix IX	31Ae	deleted
16	Appendix IX	31Af	17.18

1980 Code	1995 Code	1980 Code	1995 Code
31Ag	17.12	**Priority**	
31Ah	17.13		
31Ai	17.17 p.p.	43	Prin. 4
31Aj	17.14 p.p.	44a	13.1
31B	deleted	44b	13.1
32	28.1	44c	13.1 p.p.
32A	deleted	44 Note 1	deleted
32B	28.4, 28A.1	45a	deleted
		45b	23.5
		46	14.2

PUBLICATION AND USE OF CULTIVAR NAMES

Definitions

Retention & Re-use of cultivar names

1980 Code	1995 Code	1980 Code	1995 Code
33	9.2 p.p., 10.1	47	17.4
33 Note 1	9 Note 2	48	17.2, 26.2
33 Note 2	Preamble 5		
34	22.1 p.p.	**Rejection of cultivar names**	
35	Prin. 5, 11.1		
35A	deleted	49	27.1
36	22.1	50	6.1
36 Note 1	deleted	50 Note 1	6.2
37	21.1	50 Note 2	6.2
37 Note 1	21 Note 1	51	14.4
37 Note 2	21.3	51 Note 1	10.2
37A	21B.1	52	deleted
38	23.1		
38 Note 1	23.4	**CULTIVAR REGISTRATION**	
39	22.1 p.p.		
39A	22A.1, 22E.1	53	7 Note 1 & 2, 7.2
39B	22D.1	53 Note 1	7.3
39C	22G.1	54	deleted
40	deleted	55	deleted
40A	22B.1	55 Note 1	deleted
41	22.6	56	7.4
42	22.5		

MODIFICATION OF THE CODE

57	Div.III

IMPORTANT DATES IN THIS CODE

DATES UPON WHICH PARTICULAR PROVISIONS OF THE CODE BECOME EFFECTIVE

The Rules in this *Code* are retroactive, except in the following specified cases. The date on the left is that upon which each becomes effective.

1 May 1753	Art. 13.1
1 January 1900	Art. 21.3
1 January 1959	Art. 17.9
	Art. 17.14 p.p.
	Art. 17.16
	Art. 22.1
	Art. 23.1
1 January 1996	Art. 17.10
	Art. 17.13
	Art. 17.14 p.p.
	Art. 17.17
	Art. 17.18
	Art. 17.19
	Art. 17.20

INTERNATIONAL CODE OF NOMENCLATURE
FOR CULTIVATED PLANTS

PREAMBLE

1.   The disciplines of agriculture, forestry and horticulture require a precise, stable and simple system of nomenclature for use by practitioners of those disciplines in all countries, dealing on the one hand with the terms that denote taxonomic groups or units, and on the other hand with the names and epithets that distinguish individual taxonomic groups of cultivated plants, and which are the unique component of their scientific names.

The purpose of giving a name to a taxonomic group of cultivated plants is not to indicate its characters or history, but to supply a means of referring to it and to indicate its taxonomic status.

This *Code* aims at the provision of a stable method of naming taxonomic groups of cultivated plants, avoiding and rejecting the use of names and epithets which may cause error or ambiguity or throw the above disciplines into confusion.

Next in importance is the maintenance of prevailing custom and the avoidance of the useless creation of names and epithets.   Other considerations, such as absolute grammatical correctness, regularity or euphony of epithets, regard for persons etc., notwithstanding their undeniable importance, are relatively ancillary.

2.   The Principles form the basis of the system of nomenclature of cultivated plants.

3.   The detailed provisions are divided into Rules, set out in the Articles, and Recommendations.   Examples are added to the Rules and Recommendations to illustrate them.

4. The object of the Rules is to put the nomenclature of the past into order and to provide for that of the future; names and epithets contrary to a Rule cannot be maintained.

5. The Recommendations deal with subsidiary points, their object being to bring about greater uniformity and clearness, especially in future nomenclature; epithets contrary to a Recommendation may not, on that account, be rejected but, whenever possible, Recommendations should be followed.

6. The provisions regulating modification of this *Code* form its last Division.

7. The Rules and Recommendations apply to all organisms treated as cultivated plants whose origin or selection is primarily due to the intentional actions of mankind.

8. The *International code of botanical nomenclature*[1] provides for the botanical names in Latin form of taxonomic groups of plants; provisions for the designation of hybrids appear in Appendix I of that *Code*.

9. The only proper reasons for changing a name are either a more profound knowledge of the facts resulting from adequate taxonomic study or the necessity of giving up a nomenclature that is contrary to the Rules.

10. In the absence of a relevant Rule or where the consequences of Rules are doubtful, established custom is followed.

11. Translations of this *Code* are encouraged; in the event of discrepancy between any of these, the original English version is considered correct.

12. This edition of the *Code* supersedes all previous editions.

---

[1] Current Edition: Greuter, W., Barrie, F., Burdet, H. M., Chaloner, W. G., Demoulin, V., Hawksworth, D. L., Jørgensen, P. M., Nicolson, D. H., Silva, P. C., Trehane, P. & McNeill, J. (eds.). 1994. *International code of botanical nomenclature (Tokyo Code)*. Adopted by the Fifteenth International Botanical Congress, Yokohama, August-September 1993. (*Regnum vegetabile 131*). Koeltz, Königstein, Germany.

DIVISION I: PRINCIPLES

PRINCIPLE 1

A precise, stable and internationally recognized system for the naming of cultivated plants is essential for international understanding and communication. The aim of this *Code*, the *International code of nomenclature for cultivated plants* (*ICNCP*), also known as the Cultivated Plant Code, is to promote uniformity, accuracy and stability in the naming of agricultural, forestry and horticultural plants.

PRINCIPLE 2

The *International code of botanical nomenclature* (*ICBN* or Botanical Code) governs the botanical names in Latin form for both cultivated and wild plants, except for graft–chimaeras which are entirely governed by this *Code*. Distinguishable groups of cultivated plants, whose origin or selection is primarily due to the intentional actions of mankind, are to be given epithets formed according to the Rules and provisions of this *Code*.

The cultivated plants covered by this *Code* may arise by deliberate hybridization or by accidental hybridization in cultivation, by selection from existing cultivated stock, or may be a selection from variants within a wild population and maintained as a recognizable entity solely by continued propagation.

PRINCIPLE 3

The selection, preservation and publication of designations of Standards is important in stabilizing the application of cultivar epithets. Particular cultivar epithets are attached to Standards to make clear the precise application of the name and to help avoid duplication of such epithets. Although not a requirement for the establishment of a cultivar epithet, the designation of such Standards is strongly to be encouraged.

PRINCIPLE 4

The nomenclature of cultivars is based upon precedence of publication except in specified cases.

PRINCIPLE 5

Each taxonomic group of cultivated plants with a particular circumscription can only bear only one accepted epithet, the earliest that is in accordance with the Rules, except in specified cases.

PRINCIPLE 6

Cultivar and cultivar–group epithets must be universally available in all countries for use by any person to denote a particular cultivar or cultivar–group.

In some countries, plants are marketed using trade–marks. Such marks are assigned to a person or some corporate body and are not therefore universally available for any person to use; consequently, they cannot be considered cultivar epithets. The formation and use of trade–marks are not governed by this *Code*.

PRINCIPLE 7

This *Code* regulates the terminology to be used for taxonomic groups of cultivated plants and the epithets to be applied to individual taxonomic groups.

Under some national and international legislation, epithets may be denominated for cultivars using terminology peculiar to such legislation. This *Code* does not regulate the use of such terminology or the formation of such epithets but recognizes that, under such legislation, these statutory denominations take precedence over epithets formed under the provisions of this *Code*.

PRINCIPLE 8

The practice of applying trade designations as replacements for cultivar epithets is not supported by this *Code*. Such trade designations are not to be recognized as cultivar epithets.

PRINCIPLE 9

The formation and use of common, colloquial or vernacular names of plants are not regulated by this *Code*.

PRINCIPLE 10

Registration, preferably on an international basis, of cultivar and cultivar–group epithets and the publication of registers are of the greatest importance for promoting uniformity, accuracy and stability in the naming of
cultivated plants.

PRINCIPLE 11

This *Code* has no force beyond that deriving from the free assent of those
concerned with cultivated plants.   It is strongly urged that the Rules and
Recommendations of this *Code* be endorsed and applied by all those responsible for the formation and use of names of cultivated plants including those
responsible for statutory denominations.

PRINCIPLE 12

The provisions of this *Code* are retroactive unless expressly limited.

DIVISION II: RULES AND RECOMMENDATIONS

CHAPTER I: GENERAL PROVISIONS

Article 1: Relationship with the International Code of Botanical Nomenclature (*ICBN*)

1.1.   Cultivated plants are named in accordance with the *ICBN* at least to the level of genus, or else to the level of species or below, if, and only in so far as, they are identifiable with botanical taxa in these ranks. In addition, cultivated plants may be assigned to one or more cultivated plant taxa.

1.2.   Taxonomic groups of plants, including those for cultivated plants, will, in this *Code*, be referred to as *taxa* (singular: *taxon*).

Note 1. Taxonomic categories of cultivated plants, as distinguished from taxonomic categories of wild plants, have been termed *culta* (singular: *culton*). (See *Taxon* 44: 161–175. 1995)

CHAPTER II: DEFINITIONS

Article 2: The cultivar

2.1.   The cultivar is the basic taxon of cultivated plants.

2.2.   A cultivar is a taxon that has been selected for a particular attribute or combination of attributes, and that is clearly distinct, uniform and stable in its characteristics and that, when propagated by appropriate means, retains those characteristics.

Note 1. The botanical categories *varietas* (English: variety) and *forma* (English: form) are not the equivalent of cultivar and must not be so treated. (See Art. 17.16)

Note 2. The words "variety" in English, "variété" in French, "variedad" in Spanish, "variedade" in Portuguese, "varietà" in Italian, "ras" in Dutch, "Sorte" in German, "sort" in Scandinavian languages and Russian, "pinzhong" ("p'inchung") in Chinese, "hinshu" in Japanese, and corresponding terms in other languages, are sometimes used as common or vernacular equivalents to the word cultivar.

Note 3. The words "form" (in the sense of cultivated or garden form) in English, "Form" in German, "forme" in French, "forma" in Spanish, etc. are sometimes used as common or vernacular equivalents to the word cultivar.

2.3.   The words "variety" and "form" (or their equivalent in other languages) must not be used as synonyms for the word cultivar when fulfilling the Articles of this *Code*, nor in translations of this *Code*.

2.4.   Notwithstanding Art. 2.3, in certain national and international legislation or other legal conventions, the word "variety" is a statutory or otherwise legal term used to denominate a proven variant and is exactly equivalent to the word "cultivar" as defined in this *Code*.

2.5.   The use of the term "strain" or its equivalent in other languages, is not recognized under this *Code*, the term having been widely used in different senses so that its application is now confused.

2.6.   Cultivars differ in their mode of origin and reproduction and may be one of those defined according to Art. 2.7–2.17.

2.7.   Clones (which are asexually propagated from any part of a plant) may be given cultivar names.

Ex. 1.   *Asparagus officinalis* 'Calet', *Fraxinus pennsylvanica* 'Newport', *Gerbera* 'Delphi', *Salix alba* 'Lievelde', *Salix matsudana* 'Tortuosa', *Solanum tuberosum* 'Wilja', *Syringa vulgaris* 'Andenken an Ludwig Späth' and *Tulipa* 'Apeldoorn' are clonal cultivars.

2.8.   Topophysic clones (which must be asexually derived from particular parts of a plant) may be given cultivar names.

Ex. 2.   *Abies amabilis* 'Spreading Star' and *Abies koreana* 'Prostrate Beauty' were derived from lateral growth of the parent plants.

2.9.   Cyclophysic clones (which are derived from a particular phase of a plant's growth cycle) may be given cultivar names.

Ex. 3.   *Ficus binnendijkii* cultivars 'Amstel King', 'Amstel Queen' and 'Alii' all represent juvenile forms of the species with lanceolate leaves, whereas the adult form has elliptic leaves; *Chamaecyparis lawsoniana* 'Ellwoodii' was derived from juvenile cutting material; *Hedera helix* 'Arborescens' was derived from adult cutting material.

2.10.   Clones derived from aberrant growth may be given cultivar names.

Ex. 4.   *Picea abies* 'Pygmaea' is a dwarf plant derived from propagation of a witches' broom.

2.11.   Graft–chimaeras (which are composed of tissues from two or more different plants in intimate association and which originate by grafting) may be given cultivar names. (See also Art. 5.1)

Ex. 5.   + *Crataegomespilus* 'Dardarii' combines the tissues of *Crataegus monogyna* and *Mespilus germanica*; *Syringa* 'Correlata' combines the tissues of *S.* ×*chinensis* and *S. vulgaris*.

2.12.   Assemblages of individuals grown from seed derived from uncontrolled pollination may be given cultivar names when they meet the criteria laid down in Art. 2.2 and when they can be consistently distinguished by one or more characters, even though such individuals may not necessarily be genetically uniform.

Ex. 6. *Ballota nigra* 'Archer's Variety', *Delphinium* 'Astolat', *Geum* 'Lady Stratheden', *Lavatera* 'Ice Cool', *Milium effusum* 'Aureum', *Verbena hastata* 'Rosea', *Viola* 'Penny Black' are all cultivars which are propagated from seed.

Ex. 7. The seed-raised *Betula pendula* 'Vissingsø', *Hippophae rhamnoides* 'Aggertangen', *Larix kaempferi* 'Palsgård Velling', *Prunus padus* 'Sandgaard' and *Rosa carolina* 'Inda' were selected and refined from plants from known geographical sources.

Ex. 8. When seed is sown of the yellow-fruited cultivar *Viburnum opulus* 'Xanthocarpum', a proportion of the resulting seedlings are indistinguishable morphologically with yellow fruits, to which the cultivar epithet 'Xanthocarpum' also applies.

2.13. A line (resulting from repeated self-fertilization or inbreeding) may be given a cultivar name.

Ex. 9. *Beta vulgaris* 'SP6 926-0', *Helianthus annuus* 'HA 306', *Lactuca sativa* 'Kagraner Sommer', *Phaseolus vulgaris* 'Contender', *Triticum aestivum* 'Marquis' and *Zea mays* 'Wisconsin 153' are all lines.

2.14. A multiline (made up of several closely related lines) may be given a cultivar name.

Ex. 10. *Agropyron intermedium* 'Clarke', *Asparagus officinalis* 'Lucullus', *Glycine max* 'Jupiter-R', *Lotus corniculatus* 'Cree', *Macroptilium atropurpureum* 'Aztec' and *Trifolium repens* 'Star' are all multilines.

2.15. F_1 hybrids (the result of a repeatable single cross between two pure-bred lines) may be given a cultivar name.

Ex. 11. *Brassica oleracea* 'King Arthur', *Capsicum annuum* 'Delight', *Sorghum bicolor* 'Texas 610' are all F_1 hybrids.

2.16. An assemblage of plants grown from seed which is repeatedly collected from a particular provenance and that is clearly distinguishable by one or more characters (a topovariant) may be treated as a cultivar and named accordingly.

Ex. 12. If *Picea abies* seedlings from the Dutch provenance Gortel-1 are considered to be of recognizable high quality, they could be given the cultivar name 'Gortel-1'.

Ex. 13. Trials of *Eucalyptus camaldulensis* have demonstrated that populations from a number of different locations (provenances) produce fast-growing plants adapted to particular environmental conditions; provided such assemblages of plants meet the requirements of Art 2.2, they could be treated as cultivars and named after their original locations to provide names such as *E. camaldulensis* 'Petford', *E. camaldulensis* 'Silverton' or *E. camaldulensis* 'Lake Albacutya'.

2.17. An assemblage of genetically modified plants which shows new characteristics following the implantation of alien genetic material, may be treated as new cultivars and named accordingly.

Note 4. In practice, such plants, when man–made, are often marketed from a number of lines which have been genetically modified. These lines often remain in a constant state of development making their individual naming as cultivars a futile exercise. Generally, these lines have been marketed under trade–marks.

2.18. In considering whether two or more groups of cultivated plants belong to the same or different cultivars, the origin of each such group is irrelevant. All indistinguishable variants, irrespective of their origin, are treated as one cultivar (for seed–raised cultivars, see also Art. 3).

Ex. 14. The tobaccos described as MacNair 30 and NC2326 constitute only one cultivar since, although they derive their resistance to *Phytophthora parasitica* var. *nicotianae* from different wild species, it is not possible to distinguish one from the other.

Ex. 15. Some cultivars derived from branch sports of *Pittosporum* 'Garnettii' are indistinguishable and therefore belong to a single cultivar even though these sports have occurred at different times in different countries. *Pittosporum* 'Margaret Turnbull' which originated in New Zealand appears to be identical with *P.* 'John Flanagan' from Ireland; the International Registration Authority for *Pittosporum* designated *P.* 'Margaret Turnbull' as the accepted name and *P.* 'John Flanagan' as a taxonomic synonym.

Ex. 16. *Dianthus* 'William Sim' produces distinguishable mutants which, by further mutation, give rise to a range of variants which are indistinguishable from *D.* 'William Sim'.

2.19. If a change in the method of reproduction of a cultivar leads to changes in the set of characters by which it is distinguished, the plants so produced are not regarded as belonging to the same cultivar.

Ex. 17. The double–flowered cultivar *Campanula trachelium* 'Bernice' is vegetatively propagated. If grown from seed, it may produce a wide range of plants varying in height, degree of doubling and colour; such plants are not to be considered the same as, nor named as, *Campanula trachelium* 'Bernice'.

Ex. 18. *Cereus hildmannianus* 'Monstrosus' is a teratological cactus which is generally increased from cuttings. However on sowing seed, a proportion of seedlings show the same cristate condition. Whichever way propagation is carried out, the same name is to be applied to the cristate individuals that form the cultivar; the others are treated as indistinguishable parts of the species.

2.20. If a rootstock is a separate cultivar, it is to be named accordingly. Plants which are grafted on to rootstocks are named according to the name of the scion.

Ex. 19. The apple, *Malus domestica* 'James Grieve' grafted onto the rootstock known as *Malus domestica* 'M9', retains the epithet 'James Grieve' despite the dwarfing effect induced by that particular rootstock.

2.21. Plants whose characteristics are maintained solely by regular practices of cultivation (covariants) are not to be considered cultivars and may not receive separate cultivar epithets but must retain the name of the taxon from which they are derived.

Ex. 20. Apples trained as espaliers retain the same names as those which are tree–grown; topiary specimens of *Buxus sempervirens* may not receive cultivar names; bonsai plants retain the names of their taxon.

Article 3: Selections and maintenances

3.1.   A stock of a seed–raised cultivar may exhibit slight variation in its characteristics; these variants, although not sufficiently distinct, uniform or stable to warrant cultivar recognition (Art. 2.2), may be of practical importance. In such cases, individual stocks may be designated as selections or maintenances or by equivalent terms under locally prevailing statutory provisions.

3.2.   All plants of a selection or maintenance must retain the characteristics used to distinguish the cultivar from which they are derived.

Note 1.   A selection or maintenance may come to be recognized as a new cultivar and named accordingly if its characteristics are shown to meet the requirements of Art. 2.2.

Article 4: The cultivar–group

4.1.   Assemblages of two or more similar, named cultivars within a genus, species, nothogenus (hybrid genus), nothospecies (hybrid species) or other denomination class may be designated as cultivar–groups.

Ex. 1.   In *Primula*, the cultivars 'MacWatt's Blue', 'Old Irish Scented' and 'Osborne Green' are best cultivated under similar outdoor conditions and have been assembled under *Primula* Border Auricula Group. (See B. Hyatt, *Auriculas* 86. 1989)

Ex. 2.   In *Iris*, Dutch Group has been designated to include the complex of early flowering cultivars arising mainly from *I. tingitana*, *I. xiphium* var. *lusitanica* and *I. xiphium* var. *praecox*. (See *International checklist for hyacinths and miscellaneous bulbs* 301. 1991)

Ex. 3.   In apples, the cultivars 'Firmgold', 'Granny Smith', 'Peck's Pleasant' and 'Sturmer Pippin' have similar attributes and they have been designated as part of *Malus domestica* Granny Smith Group. (See UK Ministry of Agriculture, Fisheries and Food, *Index of the apple collection at the national fruit trials* 1985)

Ex. 4.   The cultivars of *Festuca rubra* have been allocated to three cultivar–groups, Hexaploid Non–creeping Group, Hexaploid Creeping Group and Octoploid Creeping Group, each with a distinct set of attributes. (See R. Duyvendak *et al.*, Rasen Turf Gazon 3: 53–62. 1981)

Note 1.   A cultivar–group may also be designated to cover an assemblage of cultivated plants which fall into botanical taxa that are no longer recognized as such or are of doubtful status and which, whilst exhibiting variation, show one or more characters that makes the assemblage of value.

Ex. 5.   *Rhododendron boothii* Mishmiense Group was based on *R. mishmiense*, a species no longer recognized as such but which nonetheless represents a recognizable component of the variation within *R. boothii* that continues to have horticultural utility. Any plant referable to the component of *R. boothii*, reflecting the traditional circumscription of *R. mishmiense*, may be allocated to this Mishmiense Group. (See The Royal Horticultural Society, *An alphabetical checklist of rhododendron species* 1981)

Note 2.   Prior to the introduction of cultivar–groups in this *Code*, authors may have used other designations such as "convar", "sort", "type" or "hybrids" as terminology equivalent to cultivar–group; such terms are to be replaced by "cultivar–group".

Ex. 6.   *Brachyglottis* Dunedin Hybrids was described (under *Senecio*) by D. L. Clarke (Bean, *Trees and shrubs hardy in the British Isles*, ed. 8, 4: 350. 1980) to cover a miscellany of hybrids derived from a number of species of doubtful taxonomic status. When used as a cultivar-group, the word "Hybrids" is to be replaced by "Group" so the name would be written as *Brachyglottis* Dunedin Group.

Note 3.   An assemblage which is derived from a cross in which both parents are known, and which has one or more shared characters that makes recognition of value, may also be designated as a cultivar-group.

Ex. 7.   *Lilium* Celica Group has been designated to cover all the progeny of the cross between *L.* 'Pumpkin Sweet' and *L.* 'Unique' which share the characteristics given in its description in the *International lily register*, Suppl. 11: 8. 1993.

Ex. 8.   *Rhododendron* Jacqueline Group has been designated for all the progeny of the cross between *R.* 'Albatross' and *R. facetum*. (See the *International rhododendron register* 124. 1958)

Note 4.   In orchids, a particular grouping on the basis of known parentage, of which the unit is the grex (plural: greges, although often written as grexes), is in long-standing use. This use is described in *The handbook on orchid nomenclature and registration*[1].

4.2.   When used in conjunction with a cultivar epithet, a cultivar-group designation is not part of that cultivar name but is added for information; it is placed within round brackets (parentheses) or square brackets next to the cultivar epithet.

Ex. 9.   *Dracaena fragrans* (Deremensis Group) 'Christianne', *Dracaena fragrans* 'Christianne' (Deremensis Group), *Dracaena fragrans* [Deremensis Group] 'Christianne' or *Dracaena fragrans* 'Christianne' [Deremensis Group].

4.3.   When used as an epithet (and so not in conjunction with a cultivar epithet) the cultivar-group is not to be placed within parentheses.

Ex. 10.   *Begonia* Elatior Group, *Brassica oleracea* Cauliflower Group, *Fagus sylvatica* Atropunicea Group and *Tulipa* Darwin Group, not *Begonia* (Elatior Group), *Brassica oleracea* (Cauliflower Group), *Fagus sylvatica* (Atropunicea Group) nor *Tulipa* (Darwin Group).

4.4.   A cultivar might be assigned to more than one cultivar-group should such designations have a practical purpose.

Ex. 11.   *Solanum tuberosum* 'Desiree' may be designated part of a Maincrop Group and a Red-skinned Group since both such designations may be practical to buyers of potatoes. It may thus be written *Solanum tuberosum* 'Desiree' (Maincrop Group) in one systematic classification or as *Solanum tuberosum* 'Desiree' (Red-skinned Group) in another, depending on the purpose of the classification used.

---

[1] Current (4th.) Edition: Greatwood, J., Hunt, P. F., Cribb, P. J., & Stewart, J. (eds), 1993. *The handbook on orchid nomenclature and registration*. The International Orchid Commission, London, U.K.

4.5.   When a cultivar-group is divided, or when two or more cultivar-groups are united, or when the circumscription of a cultivar-group is otherwise re-defined, a new epithet must be given for the resulting cultivar-group(s).

Ex. 12.   Recent breeding programmes in *Begonia* have led to the recognition of separate cultivar-groups within the existing Elatior Group. In due course these may be given new cultivar-group epithets instead of being referred to the Elatior Group as currently circumscribed.

Ex. 13.   In the 1950s, a number of *Magnolia* hybrids were developed by D. T. Gresham and these have been referred to as Gresham Hybrids or as the Gresham Group. The inclusion of these hybrids in a such a cultivar-group is unsatisfactory, the cultivar-group being merely a statement of origin with individual members not showing attributes in common. Two distinct cultivar-groups of Gresham's hybrids have, however, been recognized as Svelte Brunette Group and Buxom Nordic Blonde Group, each of which has a distinct set of characters. (See J. M. Gardiner, *Magnolias* 118–120. 1989)

Ex. 14.   *Tulipa* Dutch Breeders Group and *T.* English Breeders Group were united into a newly circumscribed *T.* Breeders Group. (See J. F. Ch. Dix, *A classified list of Tulip names* 4. 1958)

4.6.   The designation of grex is not to be applied except in the case of orchids (see Art. 4. Note 4).

Article 5: The graft-chimaera

5.1.   Graft-chimaeras are composed of tissues from two or more different plants which originate by grafting and are not hybrids. They are recognized as taxonomic groups under this *Code*. (See Art. 2.11 and 20)

Article 6: Denomination classes

6.1.   A denomination class is the taxon within which the use of a cultivar or cultivar-group epithet may not be duplicated except when the re-use of a cultivar epithet is sanctioned in accordance with Art. 26.2, or occurs as described in Art. 10.2 (but see Art. 17.5).

6.2.   A denomination class, under the provisions of this *Code*, is a single genus or nothogenus (their naming being governed by the provisions of the *ICBN*) unless a special denomination class has been determined by the International Commission for the Nomenclature of Cultivated Plants on application from an International Registration Authority or other statutory plant registering authority; the applicant is charged with publicising the resulting determination. (See Appendix IV for the list of current denomination classes that are not a single genus or nothogenus)

Note 1.   Statutory plant registering authorities sometimes denominate a class for the purposes of national or international legislation. Such denomination classes are usually used by such statutory authorities for the same purposes as denomination classes in this *Code*.

6.3.   When a denomination class is divided, or when two or more such denomination classes are united, or the limits of such a denomination class are changed in any way, the new denomination class is to be announced and published by the appropriate statutory plant registering authority or International Registration Authority.

Note 2. When a denomination class is a taxon whose nomenclature is governed under the *ICBN* and is divided, or when two or more such denomination classes are united, the rules of botanical nomenclature apply, unless a new denomination class is established by the appropriate International Registration Authority under the provisions of Art. 6.2.

Ex. 1. It has been proposed that the genera *Gaultheria* and *Pernettya* be united. If this proposal is followed, the denomination class is *Gaultheria* (Art. 11.3 of the *ICBN*) unless a different denomination class is determined under the provisions of Art. 6.2 of this *Code*.

Ex. 2. The genus *Ledum* is considered by some to be part of *Rhododendron*, but this view has been rejected by others. If the International Registration Authority for *Rhododendron* accepts the merging of these genera, it may apply to the International Commission for the Nomenclature of Cultivated Plants for a ruling on the limitations of the denomination class(es).

Article 7: International Registration Authorities

7.1.   An International Registration Authority is an organization charged with the registration of cultivar and cultivar-group epithets as provided for by this *Code* and with ensuring their establishment when this is necessary. (See Appendix II for a list of International Registration Authorities)

Note 1. A statutory plant registering authority is an organization established by national legal enactment or by international treaty, which, as part of its remit, publishes denominations for plants. (See Appendix III for a list of such statutory authorities)

7.2.   Registration for the purposes of this *Code* is the acceptance of a cultivar or cultivar-group epithet by an International Registration Authority and the inclusion of that epithet in a register with details of the date and place of its establishment when that information is available.

7.3.   Acceptance of an epithet by an International Registration Authority does not imply judgement on the distinctness of a cultivar or on its agricultural, horticultural or silvicultural merit. Likewise, acceptance of a cultivar-group epithet does not imply judgement on the usefulness of a cultivar-group.

7.4.   An International Registration Authority may also record trade designations and trade-marks if they have been known to have been used as marketing devices for cultivars or cultivar-groups. Such designations and trade-marks, if recorded, must not be given cultivar status and any trade-mark must not appear in a publication by an International Registration Authority except with the expressed permission of the trade-mark owner or his assignee.

Note 2.   The application for, and registration of, trade–marks which may be applied by their owners to cultivars and cultivar–groups is a legal process and not the concern of this *Code*.

Recommendation 7A

7A.1.   International Registration Authorities should, where possible, record the mode of origin and reproduction of a cultivar (Art. 2.6) and any other details mentioned in the protologue.

Recommendation 7B

7B.1.   Statutory plant registering authorities should only accept an epithet for registration after checking that the epithet is not already in the public domain by consulting any available lists and especially the checklists and registers of the relevant International Registration Authorities.

Article 8: Publication

8.1.   Publication is effected in accordance with Art. 21.

Article 9: Establishment

9.1.   Establishment of names and epithets is publication in accordance with Art. 22.

9.2.   An established epithet is one which is published in accordance with Art. 22 of this *Code*.

Note 1.   In previous editions of this *Code*, the term "validly published" was usually used in the same sense as "established" as defined above.

Note 2.   In this *Code*, unless otherwise indicated, the word "epithet" means an epithet that has been established, whether it is accepted or not.

Note 3.   An epithet that is established under the Rules of this *Code* might not be in conformity with the legislation in a particular political area.

Article 10: Accepted epithets

10.1.   The accepted cultivar epithet is the earliest one which must be adopted for it under the Rules.   More than one epithet may used as the accepted epithet under certain limited circumstances (Art. 25.1, 28.3)

Note 1.   An epithet that is not established cannot become an accepted epithet later unless it is sanctioned by an International Registration Authority as provided in Art. 14.3.

10.2.   If the same cultivar epithet has been applied to more than one cultivar within a denomination class by different statutory plant registration authorities, its application must be made precise by appending the name of the statutory authority as author of the epithet (Art. 24.5).

Article 11: Trade designations

11.1.   A trade designation is one which is used in place of the accepted cultivar epithet when the accepted epithet is considered unsuitable for marketing purposes. The accepted cultivar epithet is to be cited together with, or in juxtaposition to, the trade designation.

Ex. 1.   A number of cultivars of *Alstroemeria* are sold under trade designations rather than under their cultivar names. Under the provisions of this *Code*, both the cultivar epithet and the trade designation are to be given side–by–side and, if using the typographical example used in this *Code* (Rec. 31A.1), these would be written as *Alstroemeria* 'Stakrist' KRISTINA, *A.* 'Staromar' SACHA, *A.* 'Statiren' IRENA etc.

Ex. 2.   In 1988, UK Plant Breeders' Rights Grant No. 3743 was issued for a rose with the cultivar epithet 'Korlanum'. The cultivar is marketed under the trade designations SURREY, SOMMERWIND and VENTE D'ÉTÉ in different countries.

Ex. 3.   The name *Syringa vulgaris* 'Andenken an Ludwig Späth' was established in 1883 and is the accepted name for the cultivar. Due to prolonged use of the alternative designation "Ludwig Spaeth" by North American nurserymen, the International Registration Authority for *Syringa* has recorded the shortened epithet LUDWIG SPAETH as being a trade designation for the cultivar.

Note 1.   For the purposes of this *Code*, trade–marks are not to be taken as being trade designations as defined in this Article. (See Principle 6)

Recommendation 11A

11A.1.   Trade designations should be reported to the appropriate International Registration Authority. (See Art. 7.4)

Article 12: Standards

12.1.   The Standard for a cultivar is the designated herbarium specimen or equivalent element to which a cultivar epithet is attached, whether it is an accepted epithet or a nomenclatural or taxonomic synonym, and which should show the characters upon which its circumscription is based.

Recommendation 12A

12A.1.   Where practical, standards should be maintained in a Standard portfolio into which any specimens, illustrations or other information, such as a copy of the protologue of an epithet, may be assembled for reference purposes.

CHAPTER III: LIMITATION OF THE PRINCIPLE OF PRECEDENCE

Article 13: Starting points in nomenclature

13.1.　Establishment of cultivar and cultivar-group epithets in any denomination class is treated as starting with a list or publication designated for that denomination class by the International Commission for the Nomenclature of Cultivated Plants on the recommendation of the relevant International Registration Authority, or, in absence of such an authority, in consultation with appropriate organizations.　In the absence of such an approved list or publication, establishment starts with Linnaeus's *Species plantarum*, both volumes of which in this *Code*, as in the *ICBN*, are treated as having been published simultaneously on 1 May 1753.

Article 14: Conservation and retention of cultivar epithets

14.1.　Conservation of one cultivar epithet over another or others, whether or not the epithet to be conserved has been established as provided for under the Rules of this *Code*, is effected by the action and published decision of the International Commission for the Nomenclature of Cultivated Plants (in this Article referred to as "the Commission").　Notwithstanding the foregoing, conservation may not be applied to trade designations as they are defined in Art. 11.1 and Art. 28.1 Note 1.

Ex. 1.　The name-bearing type of *Viburnum macrocephalum* represents a snowball-flowered cultivar of horticultural significance which has been nomenclaturally isolated from the rest of the species by using the name *V. macrocephalum* 'Sterile' based on *V. macrocephalum* var. *sterile* of Dippel (Handb. Laubh. 1: 178. 1889).　Dippel's name, however, is a homotypic synonym of *V. macrocephalum* var. *macrocephalum* and as such, is not acceptable (illegitimate) under the Rules of the *ICBN* and therefore of this *Code* (Art. 17.3).　Notwithstanding the above, the epithet is in general use and the Commission has conserved the epithet 'Sterile' for use for this cultivar. (See Appendix V)

14.2.　When a cultivar epithet which has not been established in accordance with Art. 22.1 is in general use instead of what would be the accepted epithet for the same cultivar, the epithet in use may be proposed for conservation over the epithet that would be accepted if use and maintenance of that epithet might lead to confusion, or would result in an undesirable change in nomenclature.

Ex. 2.　The names *Dieffenbachia* 'Exotica Perfecta Compacta' and *D.* 'Compacta' refer to the same cultivar, the latter being more recent.　Since the first name is often inaccurately written, it has led to confusion with the cultivars *D.* 'Exotica' and *D.* 'Exotica Perfecta'.　Furthermore, the name is often written as *D.* 'Exotica Compacta'.　The later epithet 'Compacta' has been proposed as being the accepted epithet in order to avoid such ambiguity. (See Hetterscheid and van Vliet, Vakbl. Bloem. 42(50): 32–37. 1987)

14.3.   When there are two or more epithets in use for the same cultivar, the epithet that best preserves existing usage is to be chosen and sanctioned by the appropriate International Registration Authority without regard to any rank in which those epithets might have been established or to the principle of precedence.   Such an epithet becomes conserved as the accepted epithet by application to, and action of, the Commission.

Ex. 3.   *Chamaecyparis lawsoniana* 'Green Pillar' is a later name (1960) for *C. lawsoniana* 'Jackman's Variety' (1947) and has entered general use to the extent that use of the later epithet is sanctioned by the International Registration Authority for conifers (subject to ratification of conservation); that decision was published in the *International conifer register* 3: 89. 1992 (see Art. 14.4).

14.4.   When two or more cultivars in the same denomination class have received the same or a similarly confusing epithet (Art. 17.13), the cultivar to bear the accepted epithet is the one selected by an International Registration Authority.   The epithet becomes so fixed by direct reference to a designated Standard (see Art. 12.1).

Ex. 4.   When the genera *Azalea* and *Rhododendron* were merged, a number of cultivar epithets in *Azalea* were found to compete with the same ones in *Rhododendron*. In the event that the competing epithets represent different cultivars which are still extant, the International Registration Authority for rhododendrons has to decide which to retain, and which to reject. The authority may use the provisions of Art. 17.4 to provide a new epithet as a replacement for the one rejected, or the provisions of Art. 14.3 to sanction retention of a later epithet.

14.5.   An epithet that has been sanctioned by an International Registration Authority (Art. 14.3) or which has been otherwise proposed for conservation (Art. 14.2) may be used as the accepted epithet pending final ratification by the Commission.   An epithet that is sanctioned (Art. 14.3) is to take precedence over any other that is proposed.

Recommendation 14A

14A.1.   Authors should avoid resurrecting long–forgotten or obsolete epithets for taxa which have precedence over established epithets currently in use for the same taxa.

Ex. 1.   *Malus domestica* 'Mullins' Yellow Seedling' was introduced in 1916 and renamed 'Golden Delicious' in that year. The epithet 'Golden Delicious' is so well known that it should not be changed in favour of the original epithet.

Article 15: Precedence in cultivar–group epithets

15.1.   The epithets of cultivar–groups are not subject to the principle of precedence (but see Art. 15.2 and 19.2). A cultivar–group may be named and maintained for any practical purpose and for as long as that purpose remains.

15.2.   Except as provided for in Art. 28.2, the designation of a new cultivar–group epithet to replace another is not acceptable if the earlier epithet has already been applied and is in use for that cultivar–group.

CHAPTER IV: NOMENCLATURE OF TAXA OF CULTIVATED PLANTS

Article 16: Names of wild plants brought into cultivation

16.1.   Plants brought from the wild into cultivation retain the names that are applied to the same taxa growing in nature.

Ex. 1.   The common or European beech, *Fagus sylvatica*, carries the same scientific name in cultivation as in the wild.

16.2.   When two or more groups of plants are considered to contain the same individuals and no others, they are said to be coextensive.   If a cultivar or cultivar–group is considered to be coextensive with a taxon named under the provisions of the *ICBN*, it is to bear the same epithet.

Ex. 2.   Under the provisions of the *ICBN*, *Erica tetralix* forma *alba* applies to any group of plants considered to be of the rank of forma which includes the designated type of the epithet *alba*, a white flowered variant of *Erica tetralix*.   Adoption of the cultivar name *Erica tetralix* 'Alba' or the cultivar–group name *Erica tetralix* Alba Group when based upon forma *alba* implies coextension with the botanical taxon.

16.3.   Plants of a species or other infraspecific taxon brought into cultivation may not demonstrate the range of variation associated with that taxon in the wild and therefore cannot be said to be coextensive; if an assemblage of those individual plants has one or more attributes that make it worthwhile distinguishing, it may be given a cultivar name.

Ex. 3.   The distinctive plant grown in the Sir Harold Hillier Gardens and Arboretum as *Quercus frainetto* was the source (the ortet) for a large number of ramets distributed by the nursery associated with that arboretum and the resulting cultivar was named *Q. frainetto* 'Hungarian Crown' by A. J. Coombes in *The Hillier manual of trees and shrubs*, ed. 6, 348. 1991.

16.4.   Hybrids between taxa recognized under the *ICBN*, including, if it is so wished, those arising in cultivation, may receive names as provided in Appendix I of that *Code*.   Alternatively, under the provisions of this *Code*, cultivated plants arising through hybridization may be assigned to cultivars or cultivar–groups.

Ex. 4.   The nothospecies *Solanum* ×*procurrens* was published by A. C. Leslie (Watsonia 12: 29–32. 1978) for a hybrid between the European *S. nigrum* and the South American *S. physalifolium* which occurred on cultivated ground in the United Kingdom but which was not a result of human selection.

Ex. 5.   The plant selected from the cross between *Epimedium alpinum* and *E. pinnatum* subsp. *colchicum* was described and named by W. T. Stearn (Journ. Linn. Soc. Bot. 51: 521. 1938) as *E.* ×*warleyense*.   This clone is technically a nothosubspecies under the provisions of the *ICBN* but under this *Code* the same name may be given cultivar status and written as *Epimedium* 'Warleyense'.   If more than one cultivar from the same cross is subsequently selected, they might be collectively designated as *Epimedium* Warleyense Group, assuming that such a designation is considered useful.

Ex. 6. *Hypericum* ×*moserianum* was described and named by É. F. André for a hybrid between the eastern European *H. calycinum* and *H. patulum* from south–west China that arose in Moser's Nursery in Versailles, France. Under this *Code*, this hybrid may be written as *Hypericum* 'Moserianum' and any other cultivars with the same essential characteristics may, if it is so desired, be assigned to *H.* Moserianum Group or any other cultivar-group to which *H.* 'Moserianum' may be referred.

Ex. 7. *Hypericum* ×*inodorum* is the name for hybrids between *H. androsaemum* and *H. hircinum* which occur where the two species meet both in the wild and in cultivation. The taxon is not considered cultigenic and the nothospecific epithet is to be maintained; cultivars may be assigned to this nothospecies if desired.

Recommendation 16A

16A.1. Taxonomic groups of cultivated plants that meet the criteria of being recognized as cultivars or cultivar-groups should, except for nothogenera (which are named under the provisions of the *ICBN*), be named in accordance with the provisions of this *Code*, and not under the provisions of the *ICBN*.

Ex. 1. A hybrid between *Carmichaelia astonii* and *Notospartium glabrescens* arose under cultivation which was named ×*Carmispartium hutchinsii* and so validly published (established) under the provisions of the *ICBN* by M. J. Griffiths (The Plantsman 13(4): 243–248. 1992). Under the provisions of this Recommendation, a cultivar epithet in a modern language would have been provided (Art. 17.9) instead of a specific epithet; nevertheless the epithet *hutchinsii* stands, and, if the taxon is treated as a cultivar, this would be written ×*Carmispartium* 'Hutchinsii' under the provisions of Art. 17.3 of this *Code*.

Article 17: Names and epithets of cultivars

17.1. The full name of a cultivar is the accepted botanical name in Latin form of the taxon to which it is assigned, followed by the cultivar epithet.

Ex. 1. *Achillea* 'Martina', *Camellia* 'Shōjō-no-mai', *Cedrus libani* subsp. *atlantica* 'Mount Saint Catherine', *Galanthus* 'John Gray', *Magnolia* 'William Watson', *Pisum sativum* 'Consort', ×*Triticale* 'Siskiyou', *Zauschneria californica* 'Dublin'.

17.2. A cultivar epithet is not to be re-used within the denomination class except as provided for under Art. 26.2.

17.3. A Latin epithet at the rank of species or below which is validly published (established) and otherwise legitimate (acceptable) in conformity with the *ICBN* for a taxon subsequently reclassified as a cultivar (and which is considered, upon re-classification, to be coextensive with that taxon as described in Art. 6.2), is to be retained as a cultivar epithet.

Ex. 2. *Mahonia japonica* and *Primula denticulata* var. *cachmeriana* may be regarded as being of cultivated origin and when regarded as cultivars, may be written as *Mahonia* 'Japonica' and *P. denticulata* 'Cachmeriana'.

Note 1. Notwithstanding Art. 17.3, an autonym as provided for under the Rules of the *ICBN* is not to be retained as a cultivar epithet but if it represents a cultivated taxon, even if this is considered to be coextensive with the botanical taxon, is to be given an epithet in a modern language. (See Art. 25.2)

17.4. A cultivar epithet in a modern language must remain unaltered when the name of the taxon to which it is assigned is changed (Art. 14.4 and 30.2), unless, as may occur under Art. 6.3, the epithet is already in use in the denomination class to which it is assigned, in which case a new epithet will be required (but see also Art. 17.5).

Ex. 3. The following names all refer to the same cultivar: *Scilla hispanica* 'Rose Queen', *S. campanulata* 'Rose Queen', *S. hispanica* var. *campanulata* 'Rose Queen', *Hyacinthoides hispanica* 'Rose Queen' and *Endymion hispanicus* 'Rose Queen'.

17.5. Notwithstanding Art. 6.1, if a cultivar epithet in Latin form is repeated within a denomination class (Art. 6.2) but within different taxa, or if such an epithet is repeatedly in use for historical reasons, those cultivar epithets must be linked with the names of the taxa to which they apply.

Ex. 4. The supposed cultivars *Crocus goulimyi* 'Albus', *C. laevigatus* 'Albus', *C. sieberi* 'Albus' may not be written as *Crocus* 'Albus' and for precision must always include the respective specific epithet.

17.6. Each word of a cultivar epithet must start with an initial capital letter unless linguistic custom determines otherwise. Exceptions are words after a hyphen, conjunctions and prepositions other than those in the first word of the epithet, and in transliterated Japanese names, the particle "no".

Ex. 5. *Achimenes* 'Show-off', *Alonsoa warscewiczii* 'Peachy-keen', *Crocus chrysanthus* 'Eye-catcher', *Malus domestica* 'Beauty of Bath', *Narcissus* 'At Dawning', *Rosa* 'Pompon de Paris', *Camellia japonica* 'Ama-no-gawa'.

17.7. Cultivar status is to be indicated by enclosing the cultivar epithet within demarcating single quotation marks ('. . .') or, if preferred, within single vertical (downward) marks ('. . .'). Double quotation marks (". . .") and the abbreviations "cv." and "var." are not to be used within a name to distinguish cultivar epithets; such use is to be corrected.

Note 2. In past editions of this *Code*, the abbreviation "cv." was permitted as an alternative to the use of single quotation marks. Whilst no longer permitting the use of such a designation under this *Code*, it is recognized that many botanical gardens and other collections of plants are likely to continue to bear such an abbreviation on their plant labels until such time as those labels are replaced.

Ex. 6. *Garrya elliptica* 'James Roof' or *Garrya elliptica* 'James Roof', not *Garrya elliptica* "James Roof", *Garrya elliptica* cv. James Roof, *Garrya elliptica* var. James Roof etc.

Ex. 7. *Pinus sylvestris* 'Repens', not *Pinus sylvestris repens, Pinus sylvestris* var. Repens, or *Pinus sylvestris* cv. 'Repens'.

17.8. Cultivars of hybrid origin are not to be so indicated by use of the multiplication sign before the cultivar epithet.

Ex. 8. *Digitalis* 'Mertonensis' must not be written D. ×'Mertonensis'; *Distictis* 'Mrs Rivers', thought to be a hybrid between *D. buccinatoria* and *D. laxiflora* may not be written *Distictis* ×'Mrs Rivers'.

17.9. To be established, a new cultivar epithet published on or after 1 January 1959 must be a word or words in a modern language (except as otherwise required under Art. 17.3); Latin words or words which may be considered to be Latin, and thus are liable to cause confusion, may not be used unless they are the classical name of an ancient Roman person, or of a place.

Ex. 9. *Brachyglottis* (Dunedin Group) 'Moira Read', not *Brachyglottis* (Dunedin Group) 'Sunshine Variegata', *Hebe* 'Hagley Park', not *Hebe* 'Hagleyensis', *Sisyrinchium striatum* 'Aunt May', not *Sisyrinchium striatum* 'Variegatum'.

Ex. 10. *Brassica* 'Montana Sunshine' is not acceptable even if it representes a cultivar from the state of Montana in the U. S. A., since it could easily be miswritten or misunderstood in speech as *Brassica montana* 'Sunshine'.

Ex. 11. Bognor Regis, India, Mons and Nova Scotia, are place names of Latin origin which may be used in cultivar epithets; 'Balkan Peninsula', 'Harvard Campus', 'Museum d'Orsay', 'Queen of Africa' are acceptable epithets since, although containing words of Latin origin, they are unlikely to cause confusion; 'Cicero', 'Julius Caesar', 'Labianus' and 'Scipio' are acceptable epithets, whereas 'Leslieanus', an epithet derived from latinization of the modern name Leslie is not (unless that epithet exists under the provisions of Art. 17.3).

17.10. To be established on or after 1 January 1996, new cultivar epithets must consist of no more than 10 syllables and no more than 30 letters or characters overall, excluding spaces and the demarcating marks.

Note 3. In previous editions of this *Code*, epithets of more than three words were not established on or after 1 January 1959. Any epithet rejected under that ruling may not be now established under the provisions of Art. 17.10 of this *Code*.

17.11. To be established, cultivar epithets may not consist solely of common descriptive (adjectival) words in a modern language unless one of the words used may be considered a substantive, or unless the epithet is a recognized name of a colour.

Ex. 12. The cultivar epithets 'Blanche', 'Large', 'Large White', 'Double Red', 'Ogon' (Japanese for "golden") and 'Variegated' may not be established.

Note 4. For the purposes of this Article, the word "double" (or its equivalent in other languages) may not be taken to be a substantive.

Ex. 13. *Iris* (Dutch Group) 'Gold and Silver', *Ranunculus ficaria* 'Double Bronze', *Ranunculus* (Persian Group) 'Orange Brilliant' and *Zantedeschia* 'Velvet Cream' are established since, although their epithets consist of adjectival words, one or more of these could be taken to be a substantive.

Ex. 14. The epithets 'Indigo', 'Majestic Red', 'Royal Red' and 'Royal Blue' would be available since they are well-known names of specific colours.

17.12. To be established on or after 1 January 1996, epithets may not be so similar to an existing epithet in the denomination class that they might be confusing (parahomonyms).

Ex. 15.   Were it proposed for publication after 1 January 1996, *Ilex* 'Green Point' could not be established since it might be confused with the existing *I. crenata* 'Greenpoint'.

Ex. 16.   The epithet 'Susannah' could not be established if there was an existing 'Susanna' in the same denomination class; however 'Susanne' would not be considered a parahomonym as the difference in pronunciation is distinct.

Ex. 17.   *Erica carnea* 'Mrs D. E. Maxwell' could not be established since it may become confused with the existing *E. vagans* 'Mrs D. F. Maxwell'. However, *E. carnea* 'S. A. Maxwell' could be established.

Ex. 18.   The cultivar name *Narcissus* 'Miss Amy Johnson' could not be established since it could be confused with the existing *N.* 'Amy Johnson'.

17.13.   To be established on or after 1 January 1959, the botanical, common or vernacular name of any genus or nothogenus or, within a genus or nothogenus, the common or vernacular name of a species may not form a new cultivar epithet. After 1 January 1996 the botanical, common or vernacular name of any genus or nothogenus may not form any part of a new epithet unless that name is only part of the epithet and not the final word in the epithet. In addition, after 1 January 1996 the common or vernacular name of a species may not form any part of a new epithet in the genus or nothogenus to which it is assigned, nor may the botanical, common or vernacular name of a genus or nothogenus form any part of an epithet assigned to it.

Ex. 19.   After 1958, *Erica* 'Heather', *Hebe* 'Camellia' or *Populus* 'Eucalyptus' could not be established; *Geranium* 'Herb Robert' and *Lychnis* 'Jupiterbloem' could not be established since "Herb Robert" is an English common name for *Geranium robertianum* and "Jupiterbloem" is a Dutch common name for *Lychnis flos-jovis*.

Ex. 20.   After 1995, *Camellia* 'Rose Queen', *Dianthus* 'Rosa Zwerg' and *Rhododendron* 'Iris Prizeman' could be established, but *Camellia* 'Perfect Rose', *Rhododendron* 'Red Kalmia', *Viola* 'Darling Veronica', *Dianthus* 'Pink Sensation' and *Viola* 'Pat's Pansy', *Geranium* 'Old Herb Robert' and *Lychnis* 'Rode Jupiterbloem' could not be established.

Ex. 21.   *Cryptomeria japonica* 'Sekkan-sugi', containing the word element "sugi" which is the Japanese vernacular name for *Cryptomeria*, is established since it was published by G. Krüssmann in 1983. After 1995, no new cultivar epithets in *Cryptomeria* may contain the word or word element "sugi".

17.14.   Notwithstanding Art. 17.13, transliterated Japanese names could be established as cultivar epithets, even if they include the vernacular name of a species, upon removal of the word equating to the generic name, providing always that the resulting epithet in Latin script remains unique within the denomination class (see Art. 6.1).

Ex. 22.   "Ogon Setouchi Gibōshi" is the name used in Japan for a yellow-leaved cultivar of *Hosta pycnophylla*. The cultivar epithet 'Ogon Setouchi' could be derived from this name by removal of the word "Gibōshi" (which equates to the generic name *Hosta*) and by retention of the word "Setouchi" which is the vernacular name of the species.

17.15.   To be established on or after 1 January 1959, the words variety (or var.) and form may not be used in new cultivar epithets.   However, when var. denotes variegated, the epithet is established with the word "Variegated" written in full.

Ex. 23.   *Persicaria affinis* 'Lowndes' Variety' or *P. affinis* 'Lowndes' Var.' are not established because they were published after 1958.   The accepted cultivar name is *P. affinis* 'Donald Lowndes', the epithet being established as a replacement for "Lowndes' Variety".

Ex. 24.   *Iris chrysographes* 'Inshriach Form' is not established because it was published after 1958.   The name *Iris chrysographes* 'Inshriach' was established in its stead.

Ex. 25.   *Hebe* 'Longacre Variety' and *Hypericum* 'Rowallane Variety' are established because they were published before 1959.

Ex. 26.   *Astrantia major* 'Sunningdale Var.' must be corrected to *Astrantia major* 'Sunningdale Variegated' because, in this case, the abbreviation equates to the word variegated.

17.16.   To be established on or after 1 January 1996, the following words (or their equivalents in any language) may not be used in new cultivar epithets: cross, hybrid, grex, group, maintenance, mutant, seedling, selection, sport, or strain, or the plural form of these words.

Ex. 29.   'Norman Cross' or 'New Cross' could not be established, despite the fact they may commemorate the name of a person or of a place.   However 'Maycross' could be established as "cross" is only part of the word.

17.17.   To be established on or after 1 January 1996, the words improved and transformed, (or their equivalents in any language) may not be used in new cultivar epithets.   If a cultivar has been "improved" or "transformed" it must either be given a different cultivar epithet or be referred to a selection or maintenance (Art. 3.1).

17.18.   To be established on or after 1 January 1996, words exaggerating the merits of a cultivar, or which may become inaccurate through the introduction of new cultivars or other circumstances, may not be used in new cultivar epithets.

Ex. 30.   Had they been published after 1995, the apple *Malus domestica* 'Earliest of All', the bean, *Vicia faba* 'Longest Possible' and *Laburnum* 'Latest and Longest' could not be established.

17.19.   To be established on or after 1 January 1996, new epithets may not contain punctuation marks except for the apostrophe ('), the comma (,), a single exclamation mark (!), the hyphen (–) and period or full-stop (.).

Ex. 31.   The cultivar epithets 'Beryl, Viscountess Cowdray', 'Jeanne d'Arc', 'Oh Boy!', 'Hi-de-Hi', 'E. A. Bowles', 'Sing, Sing, Sing' and 'Westward Ho!' could be established.

Article 18: Designations of selections and maintenances

18.1.   The denominator of a selection or maintenance (Art. 3.1) does not form part of its cultivar name.

Note 1.   The use and formation of denominators for selections or maintenances is not governed by the Rules of this *Code*.

Article 19: Names and epithets of cultivar–groups

19.1.   The full name of a cultivar–group is the accepted botanical name in Latin form of the taxon to which it is assigned, followed by the cultivar-group epithet.

Ex. 1.   *Allium cepa* Shallot Group, *Dracaena* Deremensis Group, *Hydrangea* Hortensia Group, *Hydrangea* Lacecap Group, *Lilium* Oriental Group, *Malus domestica* Cox's Orange Pippin Group, *Rosa* Polyantha Group.

19.2.   Previously used cultivar–group epithets within the denomination class are not acceptable for re–use.

19.3.   A cultivar–group epithet is not enclosed in single or double quotation marks, but, when included in the full name of the taxon, is demarcated by round brackets (parentheses) or square brackets.

Ex. 2.   *Hydrangea macrophylla* (Hortensia Group) 'Ami Pasquier', *H. macrophylla* 'Ami Pasquier' (Hortensia Group), *H. macrophylla* [Hortensia Group] 'Ami Pasquier' or *H. macrophylla* 'Ami Pasquier' [Hortensia Group], not *H. macrophylla* "Hortensia Group" 'Ami Pasquier', *H. macrophylla* {Hortensia Group} 'Ami Pasquier' or *H. macrophylla* Hortensia Group 'Ami Pasquier'.

19.4.   Each word of a cultivar–group epithet, including the word Group, must start with an initial capital letter unless linguistic custom determines otherwise. Exceptions are words after a hyphen, conjunctions and prepositions other than those in the first word of the epithet, and, in transliterated Japanese names, the particle "no".

19.5.   If the word Group has to be abbreviated for any reason, the standard contraction Grp is to be employed, irrespective of translated equivalents of the word Group.

19.6.   To be established, a cultivar–group epithet must be a word or phrase of not more than three words in a modern language, unless based on an accepted cultivar epithet or other accepted epithet in Latin form used in the denomination class, to which is added the word Group or its equivalent in other modern languages.

Ex. 3.   *Abutilon* Darwinii Group, *Beta vulgaris* Detroit Globe Group, *Brassica rapa* Pak–Choi Group, *Fagus sylvatica* Purple-leaved Group, *Solanum aethiopicum* Gilo Group, *Vicia faba* Elatae Group.

Note 1. If a cultivar-group epithet is formed from a grex epithet (see Art. 4.6), the former grex epithet is used, with the addition of the word "Group", to form the cultivar-group epithet.

Ex. 4. Under the provisions of previous editions of this *Code*, *Lilium* Mount Shasta grex was designated for hybrids between *L. kelloggii* and *L. pardalinum*; as a cultivar-group this would be written *Lilium* Mount Shasta Group.

19.7. To be established, cultivar-group epithets may not consist solely of common descriptive words unless they are used to qualify one or more attributes of a cultivar-group.

Ex. 5. The cultivar-group epithets Large Group, White Group, Double Red Group or Purple Group could not be established, whereas the epithets Large-leaved Group, White-flowered Group, Double Red-flowered Group or Purple-leaved Group could be established.

19.8. To be established, a cultivar-group epithet may not be a later parahomonym or other confusable designation.

Ex. 6. *Clematis* Texas Group, could not be established, since it might well be confused with the existing *Clematis* Texensis Group.

19.9. To be established, a cultivar-group epithet may not contain the name (or its common or vernacular equivalent) of the genus or nothogenus to which that cultivar-group is assigned.

Ex. 7. *Epilobium* Willowherb Group or *Erica* White-flowered Heather Group may not be established, whereas *Dianthus* Malmaison Carnation Group and *Rhododendron* Kurume Azalea Group could be established.

19.10. To be established, cultivar-group epithets may not contain the following words (or their equivalents in any language): cross, form, grex, group (except as the final word), hybrid, mutant, seedling, selection, sport, strain, variety, the plural form of these words, or the words improved and transformed.

19.11. To be established, cultivar-group epithets may not contain words exaggerating the merits of a cultivar-group, or which may become inappropriate through the introduction of new cultivars or other circumstances.

Ex. 8. The cultivar-group names *Malus domestica* Tastiest Group, *Vicia faba* Biggest Harvest Group and *Laburnum* Most Glorious Group may not be established.

19.12. To be established, cultivar-group epithets may not contain punctuation marks, except the apostrophe ('), the comma (,), a single exclamation mark (!), the hyphen (–) and period or full-stop (.).

19.13. The hybrid (multiplication) sign (×) or the graft-chimaera (addition) sign (+) is not to be used before or within a cultivar-group epithet, and if used previously, is to be deleted.

Article 20: Names and epithets of graft–chimaeras

20.1.   Graft–chimaeras are designated either by a formula or by a botanical name in Latin form.

20.2.   The formula for a graft–chimaera is the accepted names of the component taxa arranged in alphabetical order linked by the addition sign +. In such formulae, a space must be left on each side of the addition sign.

Ex. 1.   *Crataegus + Mespilus*; *Cytisus purpureus + Laburnum anagyroides*; *Syringa ×chinensis + S. vulgaris.*

20.3.   When the component taxa of a graft–chimaera are different genera, a new name in Latin form may be formed by combining the generic names of the component genera, preceded by the addition sign. The generic name thus formed must not be the same as that of the nothogenus of the same genera. Such names of intergeneric graft–chimaeras are established by stating the accepted names of the component taxa.

Ex. 2.   + *Crataegomespilus* is the name for graft–chimaeras of *Crataegus* and *Mespilus* whereas × *Crataemespilus* is the nothogenus representing the sexual hybrid between *Crataegus* and *Mespilus.*

20.4.   When a new generic name has been formed as provided for by Art. 20.3, the name of a graft–chimaera cultivar consists of the generic name of the graft–chimaera followed by a cultivar epithet.

Ex. 3.   +*Laburnocytisus* 'Adamii' is the name for a graft–chimaera of *Cytisus purpureus* and *Laburnum anagyroides.*

Ex. 4.   + *Crataegomespilus* 'Dardarii' is the name of a graft–chimaera of *Crataegus monogyna* and *Mespilus germanica* whereas × *Crataemespilus gillotii* is the nothospecies derived from hybridization between the two species.

20.5.   Notwithstanding Art. 20.1, when the component taxa of a graft–chimaera belong to the same genus, the name of a graft–chimaera cultivar may comprise the generic name followed by a cultivar epithet.

Ex. 5.   *Syringa* 'Correlata' is the name for *Syringa* ×*chinensis + S. vulgaris.*

20.6.   Distinct graft–chimaeras may arise from the same component taxa in which case they are treated as separate cultivars and are to be named accordingly.

Ex. 6.   + *Crataegomespilus* 'Dardarii' and + *Crataegomespilus* 'Jules d'Asnières' are distinct cultivars of the graft–chimaera *Crataegus monogyna + Mespilus germanica.*

CHAPTER V: PUBLICATION AND ESTABLISHMENT

Article 21: Conditions of publication

21.1. Publication is effected under this *Code* only by distribution of printed or similarly duplicated matter (through sale, exchange, or gift) to the general public or at least to botanical, agricultural, forestry or horticultural institutions with libraries accessible to botanists, agriculturists, foresters or horticulturists generally. It is not effected by communication of new "names" at a public meeting, by the placing of "names" on labels in collections or gardens open to the public, by the issue of microform made from manuscripts, typescripts or other unpublished material, publication via electronic media, or by publication in confidential trade lists which are not made generally available.

Note 1. For the purpose of this Article, printed material is defined as a publication reproduced by any mechanical or graphical process whereby a number of identical, legible and indelible copies are made. The issue of microform, unpublished theses and non-technical newspapers is not regarded as a means of effective publication.

21.2. Publication may also be effected by indelible autograph.

Note 2. For the purposes of this Article, indelible autograph is handwritten material reproduced by some mechanical or graphical process (such as lithography, offset, or metallic etching).

21.3. Notwithstanding Art. 21.1–21.2, Chinese, Japanese and Korean books are considered published if, prior to 1 January 1900, they were copied by hand from a hand–written original.

Recommendation 21A

21A.1. Authors should avoid publishing new names and descriptions or diagnoses of new cultivated plant taxa covered by this *Code* in ephemeral printed matter of any kind, in particular that which is multiplied in restricted and uncertain numbers, where the permanence of the text may be limited, where the publication in terms of numbers of copies is not obvious, or where the printed matter is unlikely to reach the users described in Art. 21.1.

Recommendation 21B

21B.1. Copies of publications containing new cultivar or cultivar-group epithets should be sent to the appropriate International Registration Authorities and to suitable libraries.

Article 22: Conditions of establishment

22.1.   In order to be established, an epithet of a taxon of cultivated plants must (a) be published (Art. 21.1–21.3) on or after the starting–point date for the relevant denomination class (Art. 13.1), (b) appear in a dated publication (Art. 23.1), (c) have a form which complies with the provisions of Art. 17.9–17.20, Art. 19.6–19.12 or Art. 20.3 and (d), with the exception of intergeneric graft–chimaeras (Art. 20.3) and orchid greges (grexes) (Art. 4. Note 4), if published after 1 January 1959, be accompanied either by a description or diagnosis, or by reference to a previously published description or diagnosis.

22.2.   Notwithstanding the provisions of Art. 22.1, an epithet of a taxon of cultivated plants is established if it is accepted as a denomination in a register maintained by a statutory plant registering authority.

Note 1.   In the rules of some statutory plant registering authorities, the word "denomination" is used to mean "a given name", and equates to the word "epithet" as used in this *Code*.

Ex. 1.   *Epipremnum pinnatum* 'Golden Pothos' was accepted in the Netherlands Register of Varieties (registration number 10504, 11 October 1990). Despite the fact that it contravenes the Rules of this *Code* (a) in using the name of the genus *Pothos* and (b) in using a common name for the similar cultivar *E. pinnatum* 'Aureum' (both in contravention of Art. 17.13), the epithet 'Golden Pothos' is established under the provisions of this *Code*.

22.3.   An epithet which has not been accepted by the author who publishes it or one which has only been provisionally listed by a statutory plant registering authority is not regarded as being established.

22.4.   Correction of the original spelling of a cultivar or cultivar–group epithet (Art. 29.2) does not affect its date of establishment.

22.5.   A new cultivar epithet is not established if the cultivar of which it purports to be the epithet did not exist at the time of publication.

22.6.   A new cultivar epithet is not established if its publication is against the expressed wish of its originator or his assignee, unless the originator or his assignee had knowingly distributed that cultivar without a proposed cultivar epithet.

Ex. 2.   *Coprosma* 'The Shiner' was originally published with the epithet 'Shiner' against the wishes of its originator who had distributed the plant with his preferred name. The full name was re-instated and established by L. J. Metcalf in *The cultivation of New Zealand trees and shrubs* ed. 3, 96. 1987.

22.7.   When two or more cultivar epithets are simultaneously proposed by the same author for what the author considers to be the same cultivar, none is established.

Recommendation 22A

22A.1. The description or diagnosis of any new taxon of cultivated plants should mention the features by which the taxon differs from its allies.

Recommendation 22B

22B.1. If a description or diagnosis is in a language not using a Latin script, a translation or summary of that diagnosis or description in Latin script should be provided.

Recommendation 22C

22C.1. Reference to a previously published diagnosis or description should, whenever possible, include (a) the author's name, (b) the full title of the publication, (c) the page number on which the description or diagnosis appears and (d) the date of publication.

Recommendation 22D

22D.1. For new cultivar epithets, wherever possible, an illustration should be provided as part of the protologue.

Recommendation 22E

22E.1. When publishing a new cultivar epithet, the parentage and history of the cultivar, the etymology of the cultivar epithet and the names of the originator, nominant and introducer should be stated where known. Unless evident, the intended method of propagation should also be stated.

Recommendation 22F

22F.1. If an International Registration Authority exists for the cultivar, a copy of the publication containing the new cultivar epithet should be sent to that authority by the author of that name.

Recommendation 22G

22G.1. A specimen of the cultivar, either living or dried, should be sent to the appropriate International Registration Authority or to a public herbarium which specializes in the collection and maintenance of Standard portfolios, along with any coloured photographical or other illustrative material.

Recommendation 22H

22H.1. If possible, the institution or other organization maintaining the Standard portfolio for the cultivar should be cited when publishing a description or diagnosis of the cultivar.

Article 23: Date of publication and establishment

23.1.   On or after 1 January 1959, publication is deemed to be effected only if the publication containing the new cultivar or cultivar–group epithet is clearly dated at least to the year.

23.2.   The date of publication is the date on which publication as defined in Art. 21.1 took place.   In the absence of proof of some other date, such as the date on which the publisher or his agent delivers printed matter to one of the usual carriers for distribution to the public, the one appearing in the publication itself must be accepted.

23.3.   When separates from periodicals or other works placed on sale are issued in advance, the date on the separate is taken as being the date of publication unless there is evidence that it is erroneous.

23.4.   In cases where a dated catalogue, or other dated publication, covers more than one year, or spans a period covering part of two years, in the absence of evidence to the contrary, the earlier date given is to be taken as being the date of publication.

Ex. 1.   A dated trade catalogue covering the period Autumn 1993 to Spring 1994 is treated as having been published in Autumn 1993.

23.5.   For the purposes of determining precedence, the date of an epithet of a taxon of cultivated plants is that of its establishment.   When the various conditions for establishment are not fulfilled simultaneously, the date of the epithet is that on which the final condition was fulfilled.

Article 24: Citation of authors' names

24.1.   Except as provided in Art. 10.2, it is not necessary to cite the name of the author who has established a name of a taxon of cultivated plants under the provisions of this *Code*.

24.2.   If citation of the author of a cultivar or cultivar–group epithet is desirable this may be placed after the name, in which case the name of the author attributed with its establishment should be cited without regard to the original taxonomic position of that epithet.

Ex. 1.   The citation *Larix decidua* 'Cervicornis' Beissner may be used even though the cultivar was originally published as *L. europaea* var. *cervicornis* by Beissner, Handb. Nadelholzk. ed. 1. 328. 1891.

24.3.   If a cultivar or cultivar–group epithet is transliterated or emended as provided for in Art. 29.4, the resulting epithet is regarded as having the same author and date of publication as the original epithet.

24.4.   When an epithet is established by an International Registration Authority, the author of the name is not to be taken as that of the Registrar who publishes the epithet, but is to be taken as being the registrant of that epithet.

Ex. 2.   Narcissus 'Tidebrook' was established in the International Daffodil Register, 20th Supplement in 1994. The author of the epithet is N. A. Burr, cited as the registrant of the epithet in that publication.

24.5.   The author of an epithet denominated under the regulations of a statutory plant registering authority is not the originator or introducer of the plant, but is the denominating authority.

Ex. 3.   Ilex 'Blue Angel' and *Ilex* 'Mesog' may be written as *Ilex* 'Blue Angel' USPP 3662 and *Ilex* 'Mesog' USPP 4878 respectively, the issuing authority of the plant patent, together with its plant patent number, being the author of the epithet, not the originator of the plant (Kathleen K. Meserve) nor her assignee (The Conard–Pyle Company).

Ex. 4.   The full name and author of the following waxflower cultivar is *Chamelaucium uncinatum* 'Lady Jennifer' APBR 359 as the epithet was established in the official journal of the Australian Plant Breeders' Rights authority. (Plant Varieties Journal 7(4): 38. 1994)

Recommendation 24A

24A.1.   If the author's name is cited after the name of a taxon of cultivated plants, the provisions of Art. 46, 47 and 50 of the *ICBN*, including those for author abbreviations if desired, should be employed wherever practical.

Recommendation 24B

24B.1.   When citing a statutory plant registering authority as author of a name, the abbreviation for the authority as indicated in Appendix III should be used.

CHAPTER VI: CHOICE, RE–USE AND REJECTION OF EPITHETS

Article 25: Choice of epithets

25.1.   If a specific or infraspecific epithet is no longer thought to represent a species or infraspecific taxon, or when one taxon is made a heterotypic or homotypic synonym of another, in order to avoid a disadvantageous change of name, the subsumed epithet may be incorporated into an existing cultivar epithet, provided that the latter is also in Latin form. Such action does not effect any considerations of author citation (Art. 24.2).

Ex. 1.   If *Hosta fortunei* is not considered as a recognized species, *H. fortunei* 'Albopicta' may be written as *Hosta* 'Fortunei Albopicta' to preserve the name of the cultivar and to make make clear that the emended name refers to the same cultivar.

Ex. 2.   When *Cedrus atlantica* is made a heterotypic synonym of *C. libani*, the cultivar *C. atlantica* 'Aurea', originally attributed to A. H. Kent in 1900, may have its name changed to *C. libani* 'Atlantica Aurea' Kent, to avoid confusion with *C. libani* 'Aurea' of A. Sénéclauze published in 1868.

Note 1.   If, however, it is wished to use cultivar-groups for taxa no longer recognized as such (Art. 4. Note 1) the epithet may be used in a cultivar-group epithet.

Ex. 3.   Example 1 above would then be written as *Hosta* (Fortunei Group) 'Albopicta', there being no other epithet 'Albopicta' in the genus *Hosta* (Art. 6.1).

25.2.   When the name-bearing type of a name of a taxon validly publish-ed (established) under the provisions of the *ICBN* is considered to represent a cultivar, the name of that taxon remains unchanged but the cultivar element, if extant, is to be given a new cultivar epithet to distinguish that element from the rest of the taxon.

Ex. 4.   The name-bearing type of *Hosta sieboldii* represents a variegated cultivar of horticul-tural importance which is only partly coextensive (Art. 16.3) with the taxon. In the absence of an available epithet to encompass the cultivar, the name *Hosta sieboldii* 'Paxton's Original' was established by P. Trehane (Hortax News 1: 9. 1993).

Ex.5.   *Cotoneaster watereri* was validly published (established) (Gard. Chron. 83: 44. 1928) for a clone of cultivated origin. Under the provisions of the *ICBN* this has been treated as a hybrid and written as *Cotoneaster* ×*watereri* to which name other cultivars have been assigned. The epithet 'John Waterer' was established by C. D. Brickell (Journ. Roy. Hort. Soc. 94: 461. 1969) and this distinguishes the original cultivar from others of the same parentage.

25.3.   When two or more epithets have been established in Latin form and where, due to continued varied and competing use, it is uncertain which epithet best preserves existing usage (Art. 14.3), the earliest est-ablished epithet is to be accepted, regardless of original rank.

Ex. 6.   Under the *ICBN*, the accepted name of purple beeches at *varietas* rank is *Fagus sylvatica* var. *atropunicea* published by Weston in 1770 but at *forma* rank it is *F. sylvatica* f. *atropurpurea* published by Kirchner in 1864. Both names have been widely used. As a cultivar, the name is *F. sylvatica* 'Atropunicea'.

Article 26: Re-use of epithets

26.1.   The epithet of a cultivar or cultivar-group must not be re-used later within the same denomination class for any other cultivar or cult-ivar-group, except where a cultivar epithet is used as the basionym for the cultivar-group to which that cultivar belongs, unless re-use of a cultivar epithet is sanctioned by an International Registration Authority.

26.2.   An International Registration Authority may sanction re-use of a cultivar epithet only if that registration authority is satisfied that the original cultivar (a) is no longer in cultivation and (b) has ceased to exist as breed-ing material and (c) may not be found in a gene or seed bank and (d) is not a known component in the pedigree of other cultivars and (e) the epithet has rarely been used in publications.

26.3. When the re-use of a cultivar epithet has been sanctioned by an International Registration Authority, the sanctioned epithet and the original epithet when used, must always be accompanied by the date of publication immediately after the epithet to give precision to that epithet.

26.4. An epithet whose re-use has once been sanctioned by an International Registration Authority, may never be re-used again.

Article 27: Rejection of epithets and names

27.1 A cultivar or cultivar-group epithet, or the name of a graft-chimaera, is to be rejected if it is contrary to the Articles of this *Code*.

27.2. Notwithstanding Art. 27.1, a cultivar or cultivar-group epithet accepted and published by a statutory plant registering authority, even if using alternative terminology (Art. 2. Note 4), must not be rejected under the provisions of this *Code* (see Art. 22.2).

Ex. 1. The name of the rose 'Ausburton' was published by the United States Patent Office under Patent Number 8,838 dated July 19th, 1994 with the applicant's declaration that it was a "new and distinct variety". Although the word cultivar does not appear in the publication of grant of patent, (Art. 2.4) the cultivar name is to be taken as 'Ausburton' and may not be rejected.

27.3. A cultivar or cultivar-group epithet must not be rejected merely because (a) it is inappropriate (but see Art. 17.18) or insulting, (b) another one is preferable, or (c) it has lost its original meaning.

Ex. 2. The epithet of *Fuchsia* 'Jubelteen', meaning "upturned toes" in Dutch, may not be rejected simply because it may be thought to refer to a physical handicap.

Ex. 3. *Fraxinus ornus* 'Paus Johannes-Paulus II' was changed to *F. ornus* 'Obelisk' on the grounds that the name of a Pope should not be used in a plant name. Although some might find the original epithet sacrilegious, the epithet 'Obelisk' is treated as a superfluous cultivar epithet and is to be rejected. (Alternatively it might be considered as a trade designation as provided for under Art. 11.1).

CHAPTER VII: TRANSLATION AND TRANSLITERATION

Article 28: Translation and transliteration of epithets

28.1. When a cultivar epithet appears in a publication using a different language to that of its original publication but with the same alphabet, the epithet must not be translated.

Ex. 1. The cultivar epithet for the kohlrabi *Brassica oleracea* 'Nichtschiessender' may not be translated.

Note 1. When, for marketing reasons, a cultivar epithet has been translated into a different language, the translated epithet is treated as a trade designation (Art. 11.1).

Ex. 2.   For marketing purposes, *Hibiscus syriacus* 'L'Oiseau Bleu' might have *H. syriacus* Blue Bird as a trade designation; *Clematis* 'Blekitny Aniol' might be rendered as *C.* Blue Angel and *Cucumis sativus* 'Noas Treib' as *C. sativus* Noa's Forcing.

28.2.   When a cultivar–group epithet is used in a publication using a different language to that of its original publication but with the same alphabet, the epithet may be translated.

Ex. 3.   Purple-leaved cultivars of beech may be assigned to the cultivar–groups *Fagus sylvatica* Purple–leaved Group (in English), *F. sylvatica* Groupe à Feuilles Pourpres (in French), *F. sylvatica* Purpurblätterige Gruppe (in German), *F. sylvatica* Gruppo con Foglie Purpuree (in Italian); alternatively they may have been referred to *F. sylvatica* Atropunicea Group based on the cultivar *F. sylvatica* 'Atropunicea' (see Art. 25 Ex. 5).

28.3.   When a cultivar or cultivar–group epithet in a modern language needs to be rendered in another script, it may be transliterated.

28.4.   Transliteration into a script employing the Latin alphabet must be in accord with the following international standards: pinyin for Chinese, Hepburn for Japanese and McCune–Reischauer for Korean.

28.5.   For transliteration to or from scripts other than those specified in Art. 28.4, the standards employed by the United States Library of Congress (USLC) are to be used.

28.6.   For consistency in cultivar and cultivar–group epithets, the particle "no", derived from transliterated Japanese cultivar epithets, is to be hyphenated before and after that particle.

Ex. 4.   *Acer palmatum* 'Koto–no–ito', not *A. palmatum* 'Koto No Ito' or *A. palmatum* 'Kotonoito' ; *Camellia japonica* 'Ama–no–gawa', not *C. japonica* 'Ama no Gawa' or *C. japonica* 'Amanogawa'; *Hosta* 'Haru–no–yume', not *H.* 'Haru No Yume' or *H.* 'Harunoyume'.

Recommendation 28A

28A.1.   For the purposes of effecting the provisions of Art. 28.4–28.6, the original lettering or characters used should be examined to ensure that the transliteration conforms to those provisions.

Chapter VIII: Orthography, gender and typography

Article 29: Orthography of epithets

29.1.   The orthography of words in Latin form which are used as cultivar or cultivar–group epithets should be maintained in accordance with the provisions of the *ICBN*.

29.2. An unintentional etymological error in a cultivar or cultivar–group epithet is to be corrected.

Ex. 1. The azalea *Rhododendron* 'Sherbrook', registered with the International Registration Authority for *Rhododendron* in 1983, is to be changed to 'Sherbrooke' upon realisation that the name of place after which the cultivar was named has the latter spelling.

Ex. 2. *Argyranthemum* 'Qinta White' must not be changed to 'Quinta White' as the orthography was the deliberate choice of the originator.

29.3. Diacritical marks must be retained in cultivar and cultivar–group epithets. In determining whether a character should carry a diacritical mark or not, the original publication of the epithet should be taken as definitive, unless this is known to be contrary to the intention of the author of the epithet, or the epithet originally appeared in capital letters for which diacritical marks were absent.

29.4. Cultivar and cultivar–group epithets, or parts of such epithets, must not be abbreviated (except as provided under Art. 19.5). An exception is allowed for such epithets which include words and forms of address that are normally abbreviated in accordance with national custom. The use of such abbreviations is optional and, if desired, the abbreviated word may be spelled in full in which case the two orthographical variants are treated as the same epithet.

Ex. 3. *Astilbe* 'Prof. van der Wielen' and *Cedrus libani* 'Mt St Catherine' are acceptable cultivar epithets but may be written as *A.* 'Professor van der Wielen' and *C. libani* 'Mount Saint Catherine'.

29.5. Notwithstanding Art. 29.4, a cultivar or cultivar–group epithet originally established using the initials of personal names may not have those initials spelled out in full, nor may epithets which include an acronym or other such abbreviation have the acronym or abbreviation spelled out in full.

Ex. 4. *Crocus chrysanthus* 'E. A. Bowles' may not be changed to *C. chrysanthus* 'Edward Augustus Bowles'; the epithet 'M. Charles Coëz' may not be expanded unless the "M." means Monsieur as is the case in a *Dianthus* cultivar (see Art. 29.4); the epithet 'UN Peace Keeper' may not be written as 'United Nations Peace Keeper'.

29.6. Cognomens or nick–names may not be taken to be abbreviations but are to be treated as words in their own right when forming cultivar or cultivar–group epithets or parts of such epithets, and are not nomenclatural synonyms.

Ex. 5. *Hebe* 'Amanda Cook', *Nerium oleander* 'Albert Manning', *Malus domestica* 'Ben's Red' and *Pittosporum tenuifolium* 'James Stirling' are not to be altered to *Hebe* 'Mandy Cook', *Nerium oleander* 'Bert Manning', *Malus domestica* 'Benjamin's Red' and *Pittosporum tenuifolium* 'Jim Stirling' which might be taken as being separate cultivar names.

29.7. Hyphenated cultivar and cultivar-group epithets in a modern language must not be separated subsequently into single words unless correcting a grammatical error or an error in etymology.

Ex. 6.   *Narcissus* 'Commander-in-Chief' may not be altered to *N.* 'Commander in Chief'.

Ex. 7.   *Rhododendron* 'Terra-Cotta' is to be corrected to *R.* 'Terracotta' as it has been determined that the cultivar was named after the colour of its flowers and not after the proper name Terra-Cotta.

29.8. The initial letters of second and subsequent elements of a hyphenated word in a cultivar or cultivar-group epithet (unless conjunctions or prepositions) are to be capitalised when derived from hyphenated personal names or when taken from hyphenated place names.

Ex. 8.   *Galanthus* 'Celia Blakeway-Phillips'; *Pelargonium* 'Chi-Chi' (a famous panda); *Fuchsia* 'Shangri-La'; *Hedera helix* 'Baden-Baden'; *Nymphaea* 'Gloire de Temple-sur-Lot'.

29.9. An apparent erroneous use of a hyphen in a cultivar or cultivar-group epithet may only be corrected after examination of the etymology of the words concerned. The reasons for correction must be indicated.

29.10. The hybrid (multiplication) sign (×) or the graft-chimaera (addition) sign (+) is not to be used before or within the demarcating marks of a cultivar epithet and if used previously, is to be deleted.

Recommendation 29A

29A.1.   The liberty of correcting a cultivar or cultivar-group epithet should be used with reserve, especially if the change affects the first syllable and, above all, the first letter of such an epithet.

Article 30: Gender of names and epithets

30.1. Cultivar and cultivar-group epithets, when adjectival in form and when derived from epithets in Latin form, must agree in gender with that of the genus concerned.

Ex. 1.   *Begonia* 'Elegantissima Superba'; *Malus floribunda* 'Arnoldiana'; *Hibiscus syriacus* 'Violaceus'; *Lilium* ×*maculatum* 'Sanguineum'; *Calluna vulgaris* 'Aurea'; *Dracaena* Compacta Group.

30.2. When a cultivar or cultivar-group epithet in Latin form is transferred to a genus with another gender, the gender of the epithet is changed to agree with that of the new genus.

Ex. 2.   When *Veronica virginica* (feminine) is transferred to the genus *Veronicastrum* (neuter), the cultivar name *Veronica virginica* 'Alba' is altered to *Veronicastrum virginicum* 'Album'.

30.3. When a substantive is used as part of a cultivar or cultivar–group epithet and is qualified with an adjective, the latter agrees with the gender of the substantive, not the gender of the genus concerned.

Ex. 3. A *Rhododendron* cultivar named after the male originator Reinhold Ambrosius would be *R.* 'Ambrosius Superbus' not 'Ambrosius Superbum'. However, if both the words were adjectival in form, they would agree with the gender of the genus, i.e., *R.* 'Ambrosianum Superbum'.

Article 31: Typography of names and epithets

31.1. Cultivar and cultivar–group epithets are distinguished typographically from botanical names in Latin form; for example, they are not printed in italic typeface if the widespread convention of using italics for botanical names is adopted in the work.

Ex. 1. *Aconitum napellus* subsp. *lobelianum* 'Bergfürst' and *Chamaecyparis lawsoniana* 'Silver Queen' are not to be printed as *Aconitum napellus* subsp. *lobelianum 'Bergfürst'* or *Chamaecyparis lawsoniana 'Silver Queen'*.

Recommendation 31A

31A.1. Trade designations should be distinguished typographically from cultivar epithets; for example, they may be placed in capitals following the convention used in this *Code*. (See Art. 11.1 and 28 Ex. 2.)

CHAPTER IX: STANDARDS

Article 32: Designation of Standards

32.1. The designation of a Standard (Art. 12) is made by direct reference to the Herbarium holding the Standard.

Note 1. Under the *ICBN*, new names are fixed by the application of a name–bearing type which accompanies the description of a new taxon and which is a requirement for valid publication (establishment). This name–bearing type has to be nominated so that future workers can have, within the range of variation of the taxon, a precise reference point to which the name is permanently attached. A Standard, on the other hand, is designated so as to represent the diagnostic characteristics of a cultivated plant as well as to fix its epithet.

Note 2. It should be remembered that many cultivars, including clones, may have variable characteristics within their circumscription and any reference to a Standard in determining the identity of a plant must be done with this knowledge in mind.

32.2. A Standard may be an herbarium specimen and may have been originally nominated as a voucher specimen or as the type specimen under the provisions of the *ICBN*.

Ex. 1.   When P. C. de Jong first described *Betula utilis* 'Doorenbos' (Dendroflora 23: 26. 1986) he indicated the herbarium specimen *De Jong 1205* conserved at U[1] as a voucher specimen and this may be regarded as the Standard.

Ex. 2.   The hybrid binomen *Cotoneaster* ×*watereri* (Exell in Gard. Chron. 83: 44. 1928) is based on the name–bearing type at BM[1] and this may be considered the Standard of *C.* 'John Waterer' (see also Art. 17.3 and 25. Ex. 5).

32.3.   A Standard may be an illustration designated in place of a dried specimen when diagnostic characteristics are best recognized from a suitable illustration.

Ex. 3.   The particular flower colours of *Dianthus* and *Narcissus* cultivars and those of orchids might be better observed in paintings or photographical transparencies rather than in dried herbarium specimens.

Ex. 4.   *Aesculus pavia* 'Humilis' was first described by J. Lindley as *A. pavia* var. *humilis* in S. T. Edwards's Botanical Register accompanied by plate 1018 (1826). If these are considered coextensive (Art. 16.2), and in the absence of suitable herbarium type material, this illustration might be designated as the Standard.

32.4.   A Standard may be designated even when a name–bearing type exists if the Standard is more informative.

Ex. 5.   The Standard for *Dracaena fragrans* 'Massangeana' was designated as *Hetterscheid HDR22* (conserved at WAG) by J. J. Bos *et al.* (Edinb. Journ. Bot. 49(3): 329. 1993) even though the name–bearing type for the name of the coextensive *Dracaena massangeana* hort. ex Rodigas was lectotypified in the same paper as being the plate opposite page 170, Rev. Hort. Belge. 8: 169. 1882.

32.5   The Standard for a cultivar of grain and other crops is the sample of breeder's seed deposited and maintained in nominated seed laboratories under prevailing statutory provisions.

32.6.   The designation of a Standard, including citation of the place in which it may be found, is to be published in accordance with Art. 21.1–21.3.

Ex. 6.   The Standard for *Rhus* 'Red Autumn Lace' was designated as *A.J.Coombes 166* (conserved at WSY) by Allen J. Coombes in The New Plantsman 1(2): 110. 1994.

32.7.   The Standard for a cultivar whose epithet is established under the provisions of a statutory plant registering authority is the documentation published and/or otherwise held by the issuing authority which makes the tests made to support recognition of the new cultivar. (But see Art. 32.8)

---

[1] Abbreviations as in Holmgren, P. K., Holmgren, N. H., Barnett, L. C. (eds), *Index herbariorum* ed. 8. 1990 (*Regnum vegetabile 120*) New York Botanical Garden, Bronx, New York, U.S.A.

Ex. 7.   The Standard for *Protea* 'Possum Magic', granted Plant Breeders' Rights (certificate 310) by the Australian Plant Variety Rights Office in 1994, is the sum total of the documentation and allied information associated with that cultivar, including its description (in Plant Varieties Journal 6(1): 7. 1993), held and published by that Office.

32. 8.   The Standard for a cultivar whose cultivar epithet is established under Plant Patent statutes in the United States of America is the documentation published by the issuing statutory authority on granting approval of that patent.

Ex. 8.   The Standard for *Humulus lupulus* 'H87207-2' is the illustrations and other information accompanying the grant of Patent Number 8,823 dated July 5th, 1994.

Ex. 9.   The Standard for *Achillea* 'Anblo' is the illustrations and other information accompanying the grant of Patent Number 8,828 dated July 12th, 1994.

Recommendation 32A

32A.1.   Herbarium specimens require expert preparation and raisers of new cultivars are strongly advised to donate plant material to a recognized herbarium willing to accept such material, so that suitable specimens may be prepared. It is essential that donors inform the receiving herbarium whether the cultivar is clonal or seed–raised. Herbaria may, at their own discretion, receive and maintain specimens from third parties.

Recommendation 32B

32B.1.   A Standard that is a dried specimen should be recognized by keeping it in a specially marked folder (which may form the Standard portfolio) and should be accompanied by a description outlining the distinguishing characters of the cultivar. Where appropriate, this description should also include details of parentage and the origins of the cultivar, and should cite the date and place of establishment of the epithet.

Recommendation 32C

32C.1.   Duplicates of Standards should be circulated to other collections of Standards, especially those in other countries.

Recommendation 32D

32D.1.   Standards should be accompanied by information defining the colours of important parts of a cultivar using an internationally recognized colour chart (such as that published by the Royal Horticultural Society, UK).

Recommendation 32E

32E.1.   The exact whereabouts of any living material from which a Standard was prepared should be noted with the Standard and this information must be dated as precisely as possible since some kinds of plants are frequently moved.

Recommendation 32F

32F.1.   If the originator of the cultivar, its nominant, introducer or author of its epithet is no longer available, or willing, to examine a Standard, the advice of other experts should be sought in determining the Standard for an epithet.   If possible a taxonomist, expert in the group of plants involved, should be asked to make a taxonomic determination on the Standard.

DIVISION III: PROVISIONS FOR MODIFICATION OF THIS CODE

This *Code* may only be modified by action of the IUBS (International Union of Biological Sciences) International Commission for the Nomenclature of Cultivated Plants.

APPENDIX I

NOTES FOR INTERNATIONAL REGISTRATION AUTHORITIES

APPOINTMENT OF INTERNATIONAL REGISTRATION AUTHORITIES

International Registration Authorities (IRAs), whether they represent agricultural, forestry or horticultural disciplines, are appointed by the International Society for Horticultural Science (ISHS) and are to operate within the provisions of this *Code*.

Organizations or individuals wishing to act as IRA for a particular group of plants should apply to the Chairman, ISHS Commission for Nomenclature and Registration, c/o Royal Horticultural Society, 80 Vincent Square, London SW1P 2PE, UK. Applicants should be able to demonstrate an ability to provide for both the time and costs in setting up and maintaining an International Register, as well as providing for the costs of publication of Checklists and Registers.

Additions and amendments to the directory of IRAs shown in Appendix II are published periodically in *Chronica Horticulturae* (ISHS, Englaan 1, 6703 ET Wageningen, The Netherlands) and are available on request from the ISHS Commission for Nomenclature and Registration at the above address.

Functions of an International Registration Authority

The primary functions of an IRA are:

(a)   To register cultivar and cultivar–group epithets in the denomination class for which they have accepted responsibility and to ensure their establishment.

(b)   To publish full lists of all cultivar and cultivar–group epithets in that denomination class.

(c)   To maintain records, in as great a detail as is practical, of the origin, characteristics and history of each cultivar and cultivar–group in that denomination class.

It is NOT the function of an IRA:

(a)   To conduct trials.

(b)   To judge if one cultivar or cultivar–group is more meritorious or more useful than another.

(c)   To judge distinctness of cultivars or cultivar–groups.

The following notes are intended for the guidance of International Registration Authorities. If any Registrar encounters procedural or nomenclatural points which they find difficult to resolve, these should be referred to the ISHS Commission for Nomenclature and Registration through Dr Alan C. Leslie, RHS Garden, Wisley, Woking, Surrey GU23 6QB, United Kingdom {Tel.: (44) 1483 224234, Fax: (44) 1483 211750, E-mail: acl.reg@dial.pipex.com.}

Appointment of Registrar

Each IRA should appoint a Registrar with, if necessary, an advisory committee to assist. In some cases it may be useful to the IRA to consider arranging for a local representative in other countries to act as a collecting point for new applications. Such local registrars can filter out clearly unacceptable names and can help surmount language and cultural barriers.

Registrars will need a good working knowledge of this *Code* and of the *ICBN* as well. Access to good libraries and catalogue collections will also be necessary.

Denomination class

The IRA should fix the limits of its denomination class (or classes) following consultation with, and agreement from, the International Commission for Nomenclature of Cultivated Plants[1]. The denomination class is the taxonomic unit within which no repetition of a cultivar or cultivar–group epithet will normally be permitted.

Conserved cultivar epithets

An IRA should list any cultivar epithets that are technically unacceptable yet which they have sanctioned for use and submit these to the International Commission for Nomenclature of Cultivated Plants for confirmation as epithets to be permanently conserved over any competing epithets. Once ratified as conserved, such epithets should be noted as being conserved in any Checklist or International Register published by an IRA.

Public relations

Every effort should be made to inform the relevant specialist societies, statutory plant registration authorities and growers in all countries of the existence of registration facilities. The general public should also be informed through technical, scientific and gardening magazines, periodicals and books.

Cooperation with statutory plant registration authorities

Registrars are likely to be consulted by statutory plant registration authorities enquiring whether an epithet has already been used for a plant, prior to denomination in a statutory register. An IRA must therefore be in a position to respond to such enquiries with alacrity. IRAs must, in some situations, also be in a position to formally register an objection to an epithet proposed by a statutory plant registration authority.

---

[1] c/o The Secretary to the Commission, W. L. A. Hetterscheid, Vaste Keurings Commissie, Linnaeuslaan 2a, 1431 JV Aalsmeer, The Netherlands. {Fax: (31) 297 352 205}

Preparation of a Checklist

(a)   It is recommended that as soon as possible a new IRA should compile a prelimi-
      nary Checklist of all known cultivar and cultivar–group epithets. This should be
      circulated to various specialist authorities for comment and amendment.

(b)   Following the addition of comments and amendments from the various authorities
      the Checklist should be published. This will then act as a guide to stimulate the
      submission of further information and as a tentative International Register.

Publication of an International Register

(a)   At a reasonable interval after the Checklist has been published, so that further
      corrections, amendments, omissions and additions may be included, the IRA
      should publish a comprehensive International Register.

(b)   Subsequently, at regular intervals there should be supplements to the International
      Register, incorporating newly registered cultivar and cultivar–group epithets,
      together with corrections and amendments to those previously published.

(c)   Periodically fully revised editions of the International Register should be publish-
      ed, incorporating all supplementary lists, together with any corrections and
      amendments.

Listing requirements

(a)   The International Register should list the epithets of ALL cultivars and cul-
      tivar–groups that have been named in the denomination class(es) concerned,
      whether they are still known to be in cultivation or not. If required, cultivars
      considered to be extinct may be indicated as such, but it should be borne in mind
      that it can be extremely difficult to establish whether a particular cultivar is or is
      not still in cultivation. It is especially important to remember that some names of
      cultivars and cultivar–groups, although no longer cultivated or used, will be of
      historical importance and should be cited.

      In effect, any epithet which has been published should be included, since any
      printed reference is permanent and a potential source of confusion with any other
      use of the same epithet.

(b)   The International Register should include all synonyms (including trade designa-
      tions and any translations or transliterations that have been used) and should
      clearly indicate in each case which is the accepted epithet.

(c)   The International Register should list epithets which are not acceptable for
      registration but which are nevertheless still in commercial use. These should be
      clearly indicated as being unacceptable and not registered and, where possible, the
      reason for their unacceptability should be stated.

(d)   When an epithet has been used more than once within a denomination class, each
      use should be listed in the International Register, with the appropriate description.
      Where it can be demonstrated, precedence of publication should be indicated.

(e)   The International Register should include unregistered but established epithets, accompanied by an indicative sign to show that they have not been officially registered. This action would only be necessary for epithets established but not registered after publication of the initial Register for a particular group.

(f)   The International Register may also record trade–marks if they have been known to have been applied to cultivars or cultivar-groups. Such marks may only be published in an International Register with the expressed permission of the trade–mark owner or his assignee.

Reservation of names

An IRA should exercise caution in the reservation of epithets especially if they have not yet been definitely assigned to particular plants. However, it may, for example, be useful to allow reservation during the period in which a plant is being assessed for Plant Breeders' Rights.

Reservation of epithets should in all cases only be permitted for a definite and limited period. Registrants should be advised that a reserved epithet will no longer be available if someone else happens to establish the epithet for another plant in that denomination class.

Botanical Names

International Registers are not required to list botanical names of species or other taxonomic ranks, but some Registrars have found it useful to do so. Where this is the case, specialist help should be sought to ensure that the names used are in accordance with the latest accepted botanical treatment of the genus or genera concerned.

Registration Form

Each IRA needs to produce a registration form on which it should require all applications to be made. Advice on the general layout and content of the registration form can be provided by the ISHS Commission for Nomenclature and Registration through Dr A. C. Leslie (address above), but the detailed format will vary depending on the denomination class concerned.

All such forms should request at least the following information and material:

(a)   The names and addresses of:

(i)   the **originator** (it may be necessary to distinguish the hybridizer from the person who grew a plant on to its first flowering)

(ii) the **nominant** who designated (invented or coined) the epithet

(iii) the **introducer** (it may be necessary to distinguish private distribution or exhibition from commercial introduction)

(iv) the **registrant**

In each case the relevant date(s) should also be requested (the year is usually sufficient).

(b)   If the cultivar or cultivar–group has been named and described previously, but not registered, the name of the person who originally published the epithet, together with a copy of the publication or full reference to its date and place of publication.

(c)   If the epithet to be registered is a transliteration from a language not using the Latin alphabet the original form (characters) of the epithet should also be indicated.

(d)   The parentage, when known.

(e)   Particulars of any associated patents, trade–marks or plant breeders' rights, together with the results of testing in recognised trials, if applicable.

(f)   Awards received, with dates and the name of the awarding body.

(g)   A description in a language using the Latin alphabet including, where applicable, details of colour – the RHS Colour Chart (1966, reprinted 1986 in association with the Flower Council of Holland and in 1995 by the RHS) is now a widely used standard and is strongly recommended. The IRA should try to ensure that the registrant is prompted into giving an account of characteristics that are likely to be diagnostic.

(h)   A listing of differential characters between the cultivar or cultivar–group being registered and any closely related cultivars or cultivar–groups.

(i)   An herbarium specimen and photograph, painting or drawing to be stored subsequently in a Standard portfolio housed in a recognised herbarium.

(j)   The preferred method(s) by which any new cultivar should be propagated.

(k)   An explanation of the etymology, derivation or meaning of the cultivar epithet, especially if this is not immediately apparent from other data given on the form.

(l)   Registration forms should indicate that although the IRA will ensure eventual establishment for all registered epithets, their precedence is not fixed until publication. Registrants may thus wish to publish new epithets themselves, following registration, to ensure that precedence of their epithets is not affected.

Registration Fees

The ISHS has no funds with which to support IRAs and relies upon the voluntary cooperation of individuals and organizations in maintaining a world–wide network of registration authorities. Some IRAs charge a fee for their services, but there is no requirement that they should do so. The amount charged is left to the discretion of the IRA concerned, but charges may well discourage registration and defeat the purpose of the IRA. It should be remembered that it is not possible for a non–statutory registration authority (such as an IRA) to enforce registration as the process in itself offers no legal protection over the plant or its name.

APPENDIX II

DIRECTORY OF INTERNATIONAL REGISTRATION AUTHORITIES

A number of International Registration Authorities have regional representation in particular parts of the world and these are listed below the IRA responsible for their appointment. Anyone wishing to register a cultivar or cultivar–group epithet, or who seeks information about an epithet, is strongly recommended to contact the regional representative in the first instance.

Note 1: The dates given indicate when an IRA was first appointed (this information is not complete, especially for earlier IRAs): in some cases responsibility has been subsequently transferred to a separate organization or individual, and the later has continued to gather the information.

Note 2: The fax numbers provided are international numbers: local callers do not dial the country code shown in brackets and may have to prefix the remaining number with another digit.

GENERAL SECTION

Bulbous, cormous and tuberous-rooted plants, excluding *Dahlia* Cav., *Lilium* L., and *Narcissus* L. (1955)
ROYAL GENERAL BULBGROWERS' ASSOCIATION (KAVB), Attn: Drs Johan van Scheepen, Royal General Bulbgrowers' Association, Postbus 175, NL–2180 AD Hillegom, The Netherlands. {Fax: (31) 252 51 97 14}

Hardy Herbaceous Perennials, excluding those genera or other groups for which other IRAs have been appointed (1980)
INTERNATIONALE STAUDEN UNION (ISU), Attn: Dr Josef Sieber, Institut für Stauden und Gehölze, Versuchanstalt für Gartenbau Weihenstephan, Murstrasse 22, D–85356 Freising, Germany. {Fax: (49) 816 18 36 16}

Woody Plant Genera, excluding those genera or other groups for which other IRAs have been appointed (1958)
THE AMERICAN ASSOCIATION OF BOTANICAL GARDENS AND ARBORETA, Attn: Dr Steven Clemants, Brooklyn Botanic Garden, 1000 Washington Avenue, Brooklyn, New York 11225, United States of America. {Fax: (1) 718 857 2430, E-mail: clemants@pan-ix.com}

REGIONAL SECTION

Australian Plant Genera, excluding those genera or groups for which other IRAs have been appointed (1958)
AUSTRALIAN CULTIVAR REGISTRATION AUTHORITY, Attn: Mr Iain Dawson, Australian National Botanic Gardens, GPO Box 1777, Canberra, ACT 2602, Australia. {Fax: (61) 6 250 9599, E-mail: idawson@anbg.gov.au} [NOTE: individual genera endemic to Australia do not appear in this directory]

TAXONOMIC SECTION

Abies Mill., see under Conifers
Acacia Mill. (1958)
>   Australian Cultivar Registration Authority, Attn: Mr Iain Dawson, Australian National Botanic Gardens, GPO Box 1777, Canberra, ACT 2602, Australia. {Fax: (61) 6 250 9599, E–mail: idawson@anbg.gov.au}

Achimenes Pers., see under *Gesneriaceae*
Acidanthera Hochst. (=*Gladiolus* L.), see under Bulbous plants etc.
Actinostrobus Miq., see under Conifers
Aechmea Ruiz & Pav., see under *Bromeliaceae*
Aeschynanthus Jack, see under *Gesneriaceae*
Agathis Salisb., see under Conifers
Alcantarea (E. Morren ex Mez) Harms, see under *Bromeliaceae*
Allium L., see under Bulbous plants etc.
Aloe L. (1970)
>   The South African Aloe Breeders Association, Attn: Mr A. J. Bezuidenhout, South African Aloe Breeders Assoc., P.O. Box 59904, Karen Park, 0118, Pretoria, Republic of South Africa.

Alsobia Hanst., see under *Gesneriaceae*.
Alstroemeria L., see under Bulbous plants etc.
×*Amarcrinum* Coutts (*Amaryllis* L. × *Crinum* L.), see under Bulbous plants etc.
×*Amarygia* Cif. & Giacom. (*Amaryllis* L. × *Brunsvigia* Heist.), see under Bulbous plants etc.
Amaryllis L., see under Bulbous plants etc.
Amelanchier Medik. (1980)
>   Department of Horticulture Science, University of Saskatchewan, Attn: Dr Richard St Pierre, Department of Horticulture Science, University of Saskatchewan, Saskatoon, Saskatchewan S7N 0W0, Canada.

Amentotaxus Pilg., see under Conifers
×*Anagelia* E. L. Sm. (*Ananas* Mill. × *Neoregelia* L. B. Sm.), see under *Bromeliaceae*
×*Anamea* E. L. Sm. (*Aechmea* Ruiz & Pav. × *Ananas* Mill.), see under *Bromeliaceae*
Ananas Mill., see under *Bromeliaceae*
×*Androlaechmea* Dutrie ex C. Chev. (*Aechmea* Ruiz & Pav. × *Androlepis* Brongn. ex Houllet), see under *Bromeliaceae*
Androlepis Brongn. ex Houllet, see under *Bromeliaceae*
Andromeda L. (1970)
>   The Heather Society of Great Britain, Attn: Mr A. W. Jones, Otters Court, West Camel, Somerset BA22 7QF, United Kingdom.

Anemone L. (bulbous species), see under Bulbous plants etc.
Anomatheca Ker Gawl., see under Bulbous plants etc.
Antholyza L., see under Bulbous plants etc.
Araceae Juss. excluding *Calla* L. & *Zantedeschia* Spreng. (?1980)
>   The International Aroid Society, Attn: Mr John Banta, Route 2, Box 144, Alva, Florida 33920, United States of America. [NOTE: Individual genera of *Araceae* do not appear in this directory]

Araucaria Juss., see under Conifers
Athrotaxis D. Don, see under Conifers
Aulax P. J. Bergius, see under *Proteaceae*
Austrocedrus Florin & Boutelje (=*Libocedrus* Endl.), see under Conifers
Azalea L. (=*Rhododendron* L.), see under *Rhododendron*
Babiana Ker Gawl. ex Sims, see under Bulbous plants etc.
Bamboo, see under *Poaceae*

Begonia L. (1958)
  THE AMERICAN BEGONIA SOCIETY, Attn: Mrs Carrie Karegeannes, 3916 Lake Boule-
  vard, Annandale, Virginia 22003, United States of America.
Belamcanda Adans., see under Bulbous plants etc.
Bellevalia Lapeyr., see under Bulbous plants etc.
Billbergia Thunb., see under *Bromeliaceae*
×*Billique* E. L. Sm. (*Billbergia* Thunb. × *Quesnelia* Gaudich.), see under *Bromelia-
  ceae*
×*Billmea* Lazslo ex Padilla (*Aechmea* Ruiz & Pav. × *Billbergia* Thunb.), see under
  Bromeliaceae
Blandfordia Sm., see under Bulbous plants etc.
Bloomeria Kellogg, see under Bulbous plants etc.
Bobartia L., see under Bulbous plants etc.
Bolboxalis Small (=*Oxalis* L.), see under Bulbous plants etc.
Botryanthus Kunth (=*Muscari* Mill.), see under Bulbous plants etc.
Bougainvillea Comm. ex Juss. (1966)
  INDIAN AGRICULTURAL RESEARCH INSTITUTE (IARI), Attn: Dr Brijendra Singh, Division
  of Floriculture and Landscaping, Indian Agricultural Research Institute (IARI), New
  Delhi, 110012, India.
Brabejum L., see under *Proteaceae*
Brevoortia A. Wood (=*Dichelostemma* Kunth), see under Bulbous plants etc.
Brimeura Salisb., see under Bulbous plants etc.
Brodiaea Sm., see under Bulbous plants etc.
Bromeliaceae Juss. (1980)
  THE BROMELIAD SOCIETY INTERNATIONAL., Attn: Ellen Baskerville, R Marie Selby
  Botanical Gardens, 811 South Palm Avenue, Sarasota, Florida 34236, United States
  of America. {Fax: (1) 941 951 1474} [NOTE: only selected genera and nothogenera
  of *Bromeliaceae* appear in this directory]
Bruckenthalia Rchb. (1970)
  THE HEATHER SOCIETY OF GREAT BRITAIN, Attn: Mr A. W. Jones, Otters Court, West
  Camel, Somerset BA22 7QF, United Kingdom.
Brunsvigia Heist., see under Bulbous plants etc.
Bucinellina Wiehler, see under *Gesneriaceae*.
Bulbinella Kunth, see under Bulbous plants etc.
Bulbocodium L., see under Bulbous plants etc.
Buxus L. (1966)
  THE AMERICAN BOXWOOD SOCIETY, Attn: Mr Lynn R. Batdorf, U.S. National Arb-
  oretum, 3501 New York Avenue NE, Washington, DC 20002, United States of
  America. {Fax: (1) 202 245 45 75}
Caliphruria Herb., see under Bulbous plants etc.
Calla L., see under Bulbous plants etc.
Callicore Link (=*Amaryllis* L.), see under Bulbous plants etc.
Calliprora Lindl. (=*Triteleia* Douglas ex Lindl.), see under Bulbous plants etc.
Callistephus Cass. (1958)
  INSTITUT FÜR ZIERPFLANZENBAU, Attn: Prof. Dr K. Zimmer and Mrs Renate Noack,
  Institut für Zierpflanzenbau, Herrenhäuser Strasse 2, D–30419 Hannover–Herrenhau-
  sen, Germany. {Fax: (49) 511 76 22 654}
Callitris Vent., see under Conifers
Calluna Salisb. (1970)
  THE HEATHER SOCIETY OF GREAT BRITAIN, Attn: Mr A. W. Jones, Otters Court, West
  Camel, Somerset BA22 7QF, United Kingdom.
Calocedrus Kurz, see under Conifers
Calochortus Pursh, see under Bulbous plants etc.
Camassia Lindl., see under Bulbous plants etc.

Camellia L. (1962)
    THE INTERNATIONAL CAMELLIA SOCIETY, Attn: Mr T. J. Savige, Hawksview, Wirlinga, New South Wales 2640, Australia.
    Regional Representation, Australia:
        THE AUSTRALIAN CAMELLIA SOCIETY, Attn: Mr Ray Garnett, 36 Hardinge Street, Beaumaris, Victoria 3193, Australia.
    Regional Representation, New Zealand:
        THE NEW ZEALAND CAMELLIA SOCIETY, Attn: Mrs Yvonne Cave, Seafield, RD4 Wanganui, New Zealand.
    Regional Representation, North America:
        THE AMERICAN CAMELLIA SOCIETY, Attn: Mrs Edith Mazzei, Concord, 1486 Yosemite Circle, California 94521, United States of America.
Canistrum E. Morren, see under *Bromeliaceae*
×**Canmea** R. & M. B. Foster (*Aechmea* Ruiz & Pav. × *Canistrum* E. Morren), see under *Bromeliaceae*
Canna L., see under Bulbous plants etc.
×**Canularium** B. Deut. (*Canistrum* E. Morren × *Nidularium* Lem.), see under *Bromeliaceae*
Cathaya Chun & Kuang (=*Tsuga* Carrière), see under Conifers
Cedrus Trew, see under Conifers
Cephalotaxus Siebold & Zucc. ex Endl., see under Conifers
Chaenomeles Lindl. (1962)
    THE ARNOLD ARBORETUM, Attn: Dr Stephen A. Spongberg, Arnold Arboretum, Jamaica Plain, Massachusetts 02130, United States of America. {Fax: (1) 617 524 1418, E–mail: spongberg@arnarb.harvard.edu}
Chamaecyparis Spach, see under Conifers
Chionodoxa Boiss., see under Bulbous plants etc.
×**Chionoscilla** J. Allen ex G. Nicholson (*Chionodoxa* Boiss. × *Scilla* L.), see under Bulbous plants etc.
Choretis Herb. (=*Hymenocallis* Salisb.), see under Bulbous plants etc.
Clematis L. (1986)
    THE ROYAL HORTICULTURAL SOCIETY, Attn: Miss V. A. Matthews, 7350 SW–173rd. Street, Miami, Florida 33157–4835, United States of America.
Coburgia Herb. (=*Amaryllis* L.), see under Bulbous plants etc.
Colchicum L., see under Bulbous plants etc.
Columnea L., see under *Gesneriaceae*
Conifers (1964)
    THE ROYAL HORTICULTURAL SOCIETY, Attn: Mr Piers Trehane, % Royal Horticultural Society's Garden, Wisley, Woking, Surrey GU23 6QB, United Kingdom. {Fax: (44) 1483 211750, E–mail: coneman@indhort.demon.co.uk} [NOTE: only genera and nothogenera of horticultural or silvicultural importance appear in this directory]
    Regional Representation, Australia:
        THE CONIFER SOCIETY OF AUSTRALIA, Attn: Dr Roger Spencer, Royal Botanic Gardens Melbourne, Birdwood Avenue, South Yarra, Victoria 3141, Australia. {Fax: (61) 3 9252 2350}
    Regional Representation, North America:
        THE AMERICAN CONIFER SOCIETY, Attn: Mrs Susan Martin, U.S. National Arboretum, 3501 New York Avenue NE, Washington, DC 20002, United States of America. {Fax: (1) 202 245 45 75}
Convallaria L., see under Bulbous plants etc.
×**Cooperanthes** Lancaster (=*Zephyranthes* Herb.), see under Bulbous plants etc.
Cooperia Herb. (=*Zephyranthes* Herb.), see under Bulbous plants etc.

Coprosma J. R. Forst. & G. Forst. (1970)
  THE ROYAL NEW ZEALAND INSTITUTE OF HORTICULTURE INC., Attn: Mr L. J. Metcalf, "Greenwood", Stringers Creek, RD1, Richmond, Nelson, New Zealand.
Cornus L. (1962)
  THE ARNOLD ARBORETUM, Attn: Dr Stephen A. Spongberg, Arnold Arboretum, Jamaica Plain, Massachusetts 02130, United States of America. {Fax: (1) 617 524 1418, E–mail: spongberg@arnarb.harvard.edu}
Corydalis L., see under Bulbous plants etc.
Cotoneaster L. (1966)
  THE SIR HAROLD HILLIER GARDENS AND ARBORETUM, Attn: Mr Allen Coombes, Sir Harold Hillier Gardens and Arboretum, Jermyns Lane, Ampfield, Romsey, Hampshire SO51 0QA, United Kingdom. {Fax: (44) 1794 368027}
Crinum L., see under Bulbous plants etc.
Crocosmia Planch., see under Bulbous plants etc.
Crocus L., see under Bulbous plants etc.
Cryptanthus Otto & A. Dietr., see under *Bromeliaceae*
×*Cryptananas* B. Sm. (*Ananas* Mill. × *Cryptanthus* Otto & A. Dietr.), see under *Bromeliaceae*
×*Cryptbergia* R. G. Wilson & C. Wilson (*Billbergia* Thunb. × *Cryptanthus* Otto & A. Dietr.), see under *Bromeliaceae*
×*Cryptmea* E. L. Sm. (*Aechmea* Ruiz & Pav. × *Cryptanthus* Otto & A. Dietr.), see under *Bromeliaceae*
Cryptomeria D. Don, see under Conifers
Cunninghamia R. Br., see under Conifers
×*Cupressocyparis* Dallim. (*Chamaecyparis* Spach × *Cupressus* L.), see under Conifers
Cupressus L., see under Conifers
Curtonus N. E. Br. (=*Crocosmia* Planch.), see under Bulbous plants etc.
Cyclamen L., see under Bulbous plants etc.
Cypella Herb., see under Bulbous plants etc.
Cyperaceae Juss. (1990)
  THE HARDY PLANT SOCIETY, Attn: Dr Alan C. Leslie, Monksilver, 72 Boxgrove Road, Guildford, Surrey GU1 1UD, United Kingdom. {Fax: (44) 1483 211750} [NOTE: Individual genera of *Cyperaceae* do not appear in this directory]
Cyrtanthus Aiton, see under Bulbous plants etc.
Daboecia D. Don (1970)
  THE HEATHER SOCIETY OF GREAT BRITAIN, Attn: Mr A. W. Jones, Otters Court, West Camel, Somerset BA22 7QF, United Kingdom.
Dacrycarpus (Endl.) de Laub., see under Conifers
Dacrydium Sol. ex G. Forst., see under Conifers
Dahlia Cav. (1966)
  THE ROYAL HORTICULTURAL SOCIETY, Attn: Mr David Pycraft, Royal Horticultural Society's Garden, Wisley, Woking, Surrey GU23 6QB, United Kingdom. {Fax: (44) 1483 211750}
  Regional Representation, India:
    THE DAHLIA SOCIETY OF INDIA, Attn: Mr K. Samadder, 4 Thakur Ramkrishna, Park Row, Calcutta 25, India 700 025.
  Regional Representation, The Netherlands:
    ROYAL GENERAL BULBGROWERS' ASSOCIATION (KAVB), Attn: Drs Johan van Scheepen, Royal General Bulbgrowers' Association, Postbus 175, NL–2180 AD Hillegom, The Netherlands. {Fax: (31) 252 51 97 14}
Dalbergaria Tussac, see under *Gesneriaceae*

Delphinium L., perennials only (1955)
   THE ROYAL HORTICULTURAL SOCIETY, Attn: Dr Alan C. Leslie, Royal Horticultural
   Society's Garden, Wisley, Woking, Surrey, GU23 6QB, United Kingdom. {Fax: (44)
   1483 211750, E–mail: acl.reg@dial.pipex.com.}
Deuterocohnia Mez, see under *Bromeliaceae*
Dianthus L. (1958)
   THE ROYAL HORTICULTURAL SOCIETY, Attn: Dr Alan C. Leslie, Royal Horticultural
   Society's Garden, Wisley, Woking, Surrey GU23 6QB, United Kingdom. {Fax: (44)
   1483 211750, E–mail: acl.reg@dial.pipex.com.}
      Regional Representation, North America:
         AMERICAN DIANTHUS SOCIETY, Attn: Mr R. B. Lee, P.O. Box 22232, Santa Fe,
         New Mexico 87502, United States of America.
Diastema Benth., see under *Gesneriaceae*
Diastella Salisb. ex Knight, see under *Proteaceae*
Dichelostemma Kunth, see under Bulbous plants etc.
Dierama K. Koch, see under Bulbous plants etc.
Dioscorea L., see under Bulbous plants etc.
Diselma Hook. f., see under Conifers
Dyckia Schult. f., see under *Bromeliaceae*
×*Dycktia* B. Sm. (*Dyckia* Schult. f. × *Hechtia* Klotzsch), see under *Bromeliaceae*
Elisena Herb., see under Bulbous plants etc.
Endymion Dumort. (=*Hyacinthoides* Heist. ex Fabr.), see under Bulbous plants etc.
Episcia Mart., see under *Gesneriaceae*
Eranthis Salisb., see under Bulbous plants etc.
Eremurus M. Bieb., see under Bulbous plants etc.
Erica L., (1970)
   THE HEATHER SOCIETY OF GREAT BRITAIN, Attn: Mr A. W. Jones, Otters Court, West
   Camel, Somerset BA22 7QF, United Kingdom.
Erythronium L., see under Bulbous plants etc.
Escallonia Mutis ex L. f. (1966)
   Dr Elizabeth McClintock, 1551 9th Avenue, San Francisco, California 94122, United
   States of America.
Eucharis Planch. & Linden, see under Bulbous plants etc.
Eucodonia Hanst., see under *Gesneriaceae*
Eucomis L'Hér., see under Bulbous plants etc.
Fagus L. (1962)
   THE ARNOLD ARBORETUM, Attn: Dr Stephen A. Spongberg, Arnold Arboretum,
   Jamaica Plain, Massachusetts 02130, United States of America. {Fax: (1) 617 524
   1418, E–mail: spongberg@arnarb.harvard.edu}
Faurea Harv., see under *Proteaceae*
Ferraria Burm. ex Mill., see under Bulbous plants etc.
Forsythia Vahl (1962)
   THE ARNOLD ARBORETUM, Attn: Dr Stephen A. Spongberg, Arnold Arboretum,
   Jamaica Plain, Massachusetts 02130, United States of America. {Fax: (1) 617 524
   1418, E–mail: spongberg@arnarb.harvard.edu}
Freesia Klatt, see under Bulbous plants etc.
Fritillaria L., see under Bulbous plants etc.
Fuchsia L. (1966)
   THE AMERICAN FUCHSIA SOCIETY, Attn: Mrs Delight A. Logan, 8710 South Sheridan
   Avenue, Reedley, California 93654, United States of America.
Fumaria L., see under Bulbous plants etc.
Galanthus L., see under Bulbous plants etc.
Galtonia Decne., see under Bulbous plants etc.

Gesneriaceae Dumort., excluding *Saintpaulia* H. Wendl. (1958)
> THE AMERICAN GLOXINIA AND GESNERIAD SOCIETY INC., Attn: Judy Becker, 432 Under-mountain Road, Route 41, Salisbury, Connecticut 06068, United States of America. [NOTE: only selected genera and nothogenera of *Gesneriaceae* appear in this directory]

Ginkgo L., see under Conifers
Gladiolus L., excluding species and early flowering cultivars (1970)
> THE NORTH AMERICAN GLADIOLUS COUNCIL, Attn: Mr Samuel N. Fisher, 11734 Road 33 1/2, Madera, Califonia 93638, United States of America.

Gladiolus L., species and early flowering cultivars only, see under Bulbous plants etc.
Gleditsia L. (1962)
> THE ARNOLD ARBORETUM, Attn: Dr Stephen A. Spongberg, Arnold Arboretum, Jamaica Plain, Massachusetts 02130, United States of America. {Fax: (1) 617 524 1418, E–mail: spongberg@arnarb.harvard.edu}

Gloriosa L., see under Bulbous plants etc.
×*Gloxinera* Weathers (*Gesneria* L. × *Gloxinia* L'Hér.) see under *Gesneriaceae*
Gloxinia L'Hér., see under *Gesneriaceae*
Glyptostrobus Endl., see under Conifers
Gramineae Juss., see under [nom. alt.] *Poaceae*
×*Guzlandsia* E. L. Sm. (*Guzmania* Ruiz & Pav. × *Tillandsia* L.), see under *Bromeliaceae*
Guzmania Ruiz & Pav., see under *Bromeliaceae*
×*Guzvriesea* M. B. Foster (*Guzmania* Ruiz & Pav. × *Vriesea* Lindl.), see under *Bromeliaceae*
Gynandriris Parl., see under Bulbous plants etc.
Habranthus Herb., see under Bulbous plants etc.
Halocarpus Quinn, see under Conifers
Hebe Comm. ex Juss. (1958)
> THE ROYAL NEW ZEALAND INSTITUTE OF HORTICULTURE INC., Attn: Mr L. J. Metcalf, "Greenwood", Stringers Creek, RD1, Richmond, Nelson, New Zealand.
>> Regional Representation, United Kingdom:
>>> THE HEBE SOCIETY, Attn: Mrs Jennifer Hewitt, Haygarth, Cleeton St Mary, Kidderminster, Worcestershire DY14 0QU

×*Hechcohnia* George Anderson (*Deuterocohnia* Mez × *Hechtia* Klotzsch), see under *Bromeliaceae*
Hechtia Klotzsch, see under *Bromeliaceae*
Hedera L. (1976)
> THE AMERICAN IVY SOCIETY, Attn: Dr Sabine M. Sulgrove, 2624 Centre Creek Circle, Spring Valley, Ohio 45370, United States of America.

Hemerocallis L. (1955)
> THE AMERICAN HEMEROCALLIS SOCIETY, Attn: Mr W. C. Monroe, 7015 Chandler Drive, Baton Rouge, Louisiana 70808, United States of America.

Hepatica Mill., see under Bulbous plants etc.
Heppiella Regel, see under *Gesneriaceae*
Herbertia Sweet, see under Bulbous plants etc.
Hermodactylus Mill., see under Bulbous plants etc.
Hesperantha Ker Gawl., see under Bulbous plants etc.
Hesperopeuce (Engelm.) Lemmon (=*Tsuga* Carrière), see under Conifers
Hibiscus rosa-sinensis L. (1980)
> THE AUSTRALIAN HIBISCUS SOCIETY, Attn: Mr Christopher Noble, 61 Cockatoo Court, Caboolture, Queensland 4510, Australia.

Hippeastrum Herb., see under Bulbous plants etc.
Hohenbergia Schult. f., see under *Bromeliaceae*

×*Hohentea* B. Sm. (*Hohenbergia* Schult. f. × *Portea* K. Koch), see under *Bromeliaceae*
Homeria Vent., see under Bulbous plants etc.
Hosta Tratt. (1966)
   THE AMERICAN HOSTA SOCIETY, Attn: Mr David H. Stevenson, University of Minnesota Landscape Arboretum, P.O. Box 39, 3675 Arboretum Drive, Chanhassen, Minnesota 55317, United States of America. {Fax: (1) 612 443 2521, E-mail: steve021@maroon.tc.umn.edu}
   Regional Representation, United Kingdom:
      THE BRITISH HOSTA & HEMEROCALLIS SOCIETY Attn: Mr Roger Bowden, Sticklepath, Okehampton, Devon EX20 2NN, United Kingdom {Fax: (44) 1837 840482}
Hyacinthella Schur, see under Bulbous plants etc.
Hyacinthoides Heist. ex Fabr., see under Bulbous plants etc.
Hyacinthus L., see under Bulbous plants etc.
Hydrangea L. (1958)
   Dr Elizabeth McClintock, 1551 9th Avenue, San Francisco, California 94122, United States of America.
Hymenocallis Salisb., see under Bulbous plants etc.
Ilex L. (1955)
   THE HOLLY SOCIETY OF AMERICA, Attn: Mr Gene K. Eisenbeiss, U.S. National Arboretum, 3501 New York Avenue NE, Washington, DC 20002, United States of America. {Fax: (1) 202 245 45 75}
Ionoxalis Small (=*Oxalis* L.), see under Bulbous plants etc.
Ipheion Raf., see under Bulbous plants etc.
Iris L., bulbous species, see under Bulbous plants etc.
   Regional Representation, United Kingdom:
      THE BRITISH IRIS SOCIETY, Attn: Mrs Jennifer Hewitt, Haygarth, Cleeton St Mary, Kidderminster, Worcestershire DY14 0QU
Iris L., excluding bulbous species (1955)
   THE AMERICAN IRIS SOCIETY, Attn: Mr Kay Keppel, P.O. Box 18145, Salem, Oregon 97305, United States of America.
   Regional Representation, United Kingdom:
      THE BRITISH IRIS SOCIETY, Attn: Mrs Jennifer Hewitt, Haygarth, Cleeton St Mary, Kidderminster, Worcestershire DY14 0QU
Ismene Salisb. ex Herb. (=*Hymenocallis* Salisb.), see under Bulbous plants etc.
Ixia L., see under Bulbous plants etc.
Ixiolirion Herb., see under Bulbous plants etc.
Jovibarba Opiz (?1980)
   THE SEMPERVIVUM SOCIETY, Attn: Mr Peter J. Mitchell, 11 Wingle Tye Road, Burgess Hill, West Sussex RH15 9HR, United Kingdom. {Fax: (44) 1444 236848}
Juncaceae Juss. (1990)
   THE HARDY PLANT SOCIETY, Attn: Dr Alan C. Leslie, Monksilver, 72 Boxgrove Road, Guildford, Surrey GU1 1UD, United Kingdom. {Fax: (44) 1483 211750} [NOTE: Individual genera of *Juncaceae* do not appear in this directory]
Juniperus L., see under Conifers
Kalmia L. (1978)
   Dr Richard A. Jaynes, Broken Arrow Nursery, 13 Broken Arrow Road, Hamden, Connecticut 06518, United States of America.
Keteleeria Carrière, see under Conifers
Koellikeria Regel, see under *Gesneriaceae*
Kohleria Regel, see under *Gesneriaceae*
Korolkowia Regel, see under Bulbous plants etc.
Lachenalia J. Jacq. ex Murray, see under Bulbous plants etc.

Lagerostrobus Quinn (=*Dacrydium* Sol. ex G. Forst.), see under Conifers
Lagerstroemia L. (1970)
> THE UNITED STATES NATIONAL ARBORETUM, U.S. National Arboretum, 3501 New York Avenue NE, Washington, DC 20002, United States of America. {Fax: (1) 202 245 45 75}

Lantana L. (1970)
> THE ARNOLD ARBORETUM, Attn: Dr Stephen A. Spongberg, Arnold Arboretum, Jamaica Plain, Massachusetts 02130, United States of America. {Fax: (1) 617 524 1418, E-mail: spongberg@arnarb.harvard.edu}

Lapeirousia Pourr., see under Bulbous plants etc.
Larix Mill., see under Conifers
Lepidothamnus Phil. (=*Dacrydium* Sol. ex G. Forst.), see under Conifers
Leptospermum J. R. Forst. & G. Forst. (1958)
> THE ROYAL NEW ZEALAND INSTITUTE OF HORTICULTURE INC., Attn: Mr L. J. Metcalf, "Greenwood", Stringers Creek, RD1, Richmond, Nelson, New Zealand.

Leucadendron R. Br., see under *Proteaceae*
Leucocoryne Lindl., see under Bulbous plants etc.
Leucojum L., see under Bulbous plants etc.
Leucospermum R. Br., see under *Proteaceae*
Libocedrus Endl., see under Conifers
Lilium L. (1958)
> THE ROYAL HORTICULTURAL SOCIETY, Attn: Dr Alan C. Leslie, Royal Horticultural Society's Garden, Wisley, Woking, Surrey GU23 6QB, United Kingdom. {Fax: (44) 1483 211705}
>
> Regional Representation, Australia:
>> THE AUSTRALIAN LILIUM SOCIETY, Attn: Mr R. Macgregor, 426 Belgrave Gembrook Road, Emerald, Victoria 3782, Australia.
>
> Regional Representation, Czech Republic:
>> Ing. B. Mičulka, 687 06 Velehrad 225, Czech Republic.
>
> Regional Representation, The Netherlands:
>> ROYAL GENERAL BULBGROWERS' ASSOCIATION (KAVB), Attn: Drs Johan van Scheepen, Royal General Bulbgrowers' Association, Postbus 175, NL-2180 AD Hillegom, The Netherlands. {Fax: (31) 252 51 97 14}
>
> Regional Representation, New Zealand:
>> THE NEW ZEALAND LILY SOCIETY, Attn: Mrs M. L. Lepper, P.O. Box 1394, Christchurch, New Zealand.
>
> Regional Representation, Russia:
>> Dr A. V. Otroshko, Postbox 3, Moscow 107061, Russia.

Lycoris Herb., see under Bulbous plants etc.
Magnolia L. (?1962)
> THE MAGNOLIA SOCIETY, Attn: Dorothy J. Callaway, P.O. Box 3131, Thomasville, Georgia 31799, United States of America. {Fax: (1) 912 228 5836}

Malus Mill., ornamental cultivars only (1958)
> THE ARNOLD ARBORETUM, Attn: Dr Stephen A. Spongberg, Arnold Arboretum, Jamaica Plain, Massachusetts 02130, United States of America. {Fax: (1) 617 524 1418, E-mail: spongberg@arnarb.harvard.edu}

Mangifera indica L. (Mango) (1970)
> THE INDIAN AGRICULTURAL RESEARCH INSTITUTE, Attn: Dr S. N. Pandey, Div. of Fruits & Horticultural Tech., Indian Agricultural Research Institute, New Dehli 110012, India.

Metasequoia Hu & W. C. Cheng, see under Conifers
Microbiota Kom., see under Conifers
Milla Cav., see under Bulbous plants etc.
Mimetes Salisb., see under *Proteaceae*

Monopyle Moritz ex Benth., see under *Gesneriaceae*
Moussonia Regel, see under *Gesneriaceae*
Montbretia DC. (=*Tritonia* Ker Gawl.), see under Bulbous plants etc.
Moraea Mill., see under Bulbous plants etc.
Muscari Mill., see under Bulbous plants etc.
Muscarimia Kostel. ex Losinsk., see under Bulbous plants etc.
Nageia Gaertn., see under Conifers
Narcissus L. (1955)
> THE ROYAL HORTICULTURAL SOCIETY, Attn: Mrs Sally Kington, Royal Horticultural Society, P.O. Box 313, Vincent Square, London SW1P 2PE, United Kingdom. {Fax: (44) 171 630 6060}
> Regional Representation, Australia:
>> THE NATIONAL DAFFODIL ASSOCIATION OF AUSTRALIA, Attn: Mr Tony Davis, 4 Carandini Street, Melba, ACT 2615, Australia.
> Regional Representation, Czech Republic:
>> NARCIS KLADNO, Attn: Ing. Vladimír Domský, Konevova 40, 400 00 Usti nad Labem, Czech Republic.
> Regional Representation, The Netherlands:
>> ROYAL GENERAL BULBGROWERS' ASSOCIATION (KAVB), Attn: Drs Johan van Scheepen, Royal General Bulbgrowers' Association, Postbus 175, NL-2180 AD Hillegom, The Netherlands. {Fax: (31) 252 51 97 14}
> Regional Representation, New Zealand:
>> THE NATIONAL DAFFODIL SOCIETY OF NEW ZEALAND, Attn: Mr Max Hamilton, Boyd Road, RD1, Hamilton, New Zealand.
> Regional Representation, North America:
>> THE AMERICAN DAFFODIL SOCIETY, Attn: Mrs Mary Lou Gripshover, 1686 Grey Fox Trails, Milford, Ohio 45150, United States of America. {Fax: (1) 513 248 0898}
Nectaroscordum Lindl., see under Bulbous plants etc.
Nelumbo L. (1988)
> THE INTERNATIONAL WATER LILY SOCIETY, Attn: Mr Philip R. Swindells, Vale Lodge, Ropley, Harrogate, North Yorkshire HG3 3AY, United Kingdom. {Fax: (44) 1423 568080}
Nematanthus Schrad., see under *Gesneriaceae*
×*Neobergia* E. L. Sm. (*Billbergia* Thunb. × *Neoregelia* L. B. Sm.), see under *Bromeliaceae*
Neocallitropsis Florin, see under Conifers
×*Neoistrum* B. Sm. (*Canistrum* E. Morren × *Neoregelia* L. B. Sm.), see under *Bromeliaceae*
×*Neolarium* R. & M. B. Foster (*Neoregelia* L. B. Sm. × *Nidularium* Lem.), see under *Bromeliaceae*
×*Neomea* R. & M. B. Foster (*Aechmea* Ruiz & Pav. × *Neoregelia* L. B. Sm.), see under *Bromeliaceae*
×*Neophytum* R. & M. B. Foster (*Neoregelia* L. B. Sm. × *Orthophytum* Beer), see under *Bromeliaceae*
Neoregelia L. B. Sm., see under *Bromeliaceae*
×*Neotanthus* E. L. Sm. (*Cryptanthus* Otto & A. Dietr. × *Neoregelia* L. B. Sm.), see under *Bromeliaceae*
Nerine Herb., see under Bulbous plants etc.
Nidularium Lem., see under *Bromeliaceae*
×*Nidbergia* D. Butcher (*Billbergia* Thunb. × *Nidularium* Lem.), see under *Bromeliaceae*
×*Nidumea* L. B. Sm. (*Aechmea* Ruiz & Pav. × *Nidularium* Lem.), see under *Bromeliaceae*

Niphaea Lindl., see under *Gesneriaceae*
Nothoscordum Kunth, see under Bulbous plants etc.
Nymphaea L. (1988)
   THE INTERNATIONAL WATER LILY SOCIETY, Attn: Mr Philip R. Swindells, Vale Lodge, Ropley, Harrogate, North Yorkshire HG3 3AY, United Kingdom. {Fax: (44) 1423 568080}
Orchidaceae Juss. (1972)
   THE ROYAL HORTICULTURAL SOCIETY, Attn: Mr Peter F. Hunt, P.O. Box 1072, Frome, Somerset BA11 5NY, United Kingdom. {Fax: (44) 171 630 6060 or (44) 1483 211705}
   [NOTE: Individual genera of *Orchidaceae* do not appear in this directory. Only grex names are registered: there is no registration of cultivar names for orchids at present]
Ornithogalum L., see under Bulbous plants etc.
Orothamnus Pappe ex Hook., see under *Proteaceae*
× *Ortholarium* R. & M. B. Foster (*Nidularium* Lem. × *Orthophytum* Beer), see under *Bromeliaceae*
× *Orthomea* E. L. Sm. (*Aechmea* Ruiz & Pav. × *Orthophytum* Beer), see under *Bromeliaceae*
Orthophytum Beer, see under *Bromeliaceae*
× *Orthotanthus* Garretson (*Cryptanthus* Otto & A. Dietr. × *Orthophytum* Beer), see under *Bromeliaceae*
Oxalis L., see under Bulbous plants etc.
Paeonia L. (1974)
   THE AMERICAN PEONY SOCIETY, Attn: Mrs Greta Kessenich, 250 Interlachen Road, Hopkins, Minnesota 55343, United States of America.
Pancratium L., see under Bulbous plants etc.
Parakohleria Wiehler, see under *Gesneriaceae*
Paranomus Salisb., see under *Proteaceae*
Pardanthus Ker Gawl. (=*Belamcanda* Adans.), see under Bulbous plants etc.
Pelargonium Aiton (?1970)
   THE AUSTRALIAN GERANIUM SOCIETY, Attn: Mrs J. D. Llewellyn, "Nyndee", 56 Torokina Avenue, St Ives, New South Wales 2075, Australia.
Penstemon Schmidel (?1966)
   THE AMERICAN PENSTEMON SOCIETY, Attn: Dr Dale T. Lindgren, University of Nebraska, West Central Center, Route 4, Box 46A, North Platte, Nebraska 69101, United States of America.
Petunia Juss. (1966)
   INSTITUT FÜR ZIERPFLANZENBAU, Attn: Prof. Dr K. Zimmer and Mrs Renate Noack, Institut für Zierpflanzenbau, Herrenhäuser Strasse 2, D–30419 Hannover–Herrenhausen, Germany. {Fax: (49) 511 76 22 654}
Philadelphus L. (1958)
   THE ARNOLD ARBORETUM, Attn: Dr Stephen A. Spongberg, Arnold Arboretum, Jamaica Plain, Massachusetts 02130, United States of America. {Fax: (1) 617 524 1418, E–mail: spongberg@arnarb.harvard.edu}
Phormium J. R. Forst. & G. Forst. (?1970)
   THE ROYAL NEW ZEALAND INSTITUTE OF HORTICULTURE INC., Attn: Mr L. J. Metcalf, "Greenwood", Stringers Creek, RD1, Richmond, Nelson, New Zealand.
Phyllocladus Rich. & Mirb., see under Conifers
Picea A. Dietr., see under Conifers
Pieris D. Don (1962)
   THE ARNOLD ARBORETUM, Attn: Dr Stephen A. Spongberg, Arnold Arboretum, Jamaica Plain, Massachusetts 02130, United States of America. {Fax: (1) 617 524 1418, E–mail: spongberg@arnarb.harvard.edu}
Pinus L., see under Conifers

Pitcairnia L'Hér., see under *Bromeliaceae*
Pittosporum Banks ex Sol. (?1974)
  THE ROYAL NEW ZEALAND INSTITUTE OF HORTICULTURE INC., Attn: Mr L. J. Metcalf,
  "Greenwood", Stringers Creek, RD1, Richmond, Nelson, New Zealand.
Platycladus Spach (=*Thuja* L.), see under Conifers
Plumeria L. (1980)THE PLUMERIA SOCIETY OF AMERICA INC., Attn: Mr John P. Oliver,
  P.O. Box 22791, Houston, Texas 77227-2791, United States of America.
Poaceae (R. Br.) Barnhart [nom. alt. *Gramineae* Juss.], excluding turf, silage and cereal
  crops (1990)
  THE HARDY PLANT SOCIETY, Attn: Dr Alan C. Leslie, Monksilver, 72 Boxgrove Road,
  Guildford, Surrey GU1 1UD, United Kingdom. {Fax: (44) 1483 211750} [NOTE:
  Individual genera of *Gramineae* do not appear in this directory]
Podocarpus L'Hér. ex Pers., see under Conifers
Populus L., forestry cultivars (1955)
  THE INTERNATIONAL POPLAR COMMISSION, Attn: Mr J. B. Ball, Secretary, International
  Poplar Commission, Food and Agriculture Organization of the United Nations
  (FAO), Viale delle Terme di Caracalla, 00100 Rome, Italy. {Fax: (39) 6 522 5137,
  E–mail: james.ball@fao.org}
Portea K. Koch, see under *Bromeliaceae*
×*Portemea* Ariza–Julia ex Padilla (*Aechmea* Ruiz & Pav. × *Portea* K. Koch), see
  under *Bromeliaceae*
Potentilla fruticosa L. *sensu lato* (1966)
  AGRICULTURE CANADA RESEARCH STATION, Attn: Dr Campbell G. Davidson, Agriculture
  Canada Research Station, Unit 100–101 Route 100, Morden, Manitoba R6M 1Y5,
  Canada. {Fax: (1) 204 822 6841}
Protea L., see under *Proteaceae*
Proteaceae Juss., South African genera only (1980)
  DIRECTORATE OF PLANT QUALITY CONTROL, SOUTH AFRICA DEPARTMENT OF AGRICULTURE,
  Attn: Mr M. S. Joubert, Directorate of Plant Quality Control, Department of
  Agriculture, Private Bag X258, Pretoria 0001, Republic of South Africa. {Fax: (27)
  12 319 6055}
Prumnopitys Phil., see under Conifers
Pseudolarix Gordon, see under Conifers
Pseudomuscari Garbari & Greuter, see under Bulbous plants etc.
Pseudotsuga Carrière, see under Conifers
Puschkinia Adams, see under Bulbous plants etc.
×*Puycohnia* George Anderson (*Deuterocohnia* Mez × *Puya* Molina), see under
  Bromeliaceae
Puya Molina, see under *Bromeliaceae*
Pyracantha M. Roem. (1966)
  THE UNITED STATES NATIONAL ARBORETUM, U.S. National Arboretum, 3501 New York
  Avenue NE, Washington, DC 20002, United States of America. {Fax: (1) 202 245
  45 75}
×*Quesmea* R. & M. B. Foster (*Aechmea* Ruiz & Pav. × *Quesnelia* Gaudich.), see
  under *Bromeliaceae*
Quesnelia Gaudich., see under *Bromeliaceae*
×*Quesregelia* Carrone (*Neoregelia* L. B. Sm. × *Quesnelia* Gaudich.), see under
  Bromeliaceae
Ranunculus L., bulbous species, see under Bulbous plants etc.
Rhododendron L., including *Azalea* L. (1955)
  THE ROYAL HORTICULTURAL SOCIETY, Attn: Dr Alan C. Leslie, Royal Horticultural
  Society's Garden, Wisley, Woking, Surrey GU23 6QB, United Kingdom. {Fax: (44)
  1483 211750, E–mail: acl.reg@dial.pipex.com.}

Rhododendron – continued
Regional Representation, Australia:
THE AUSTRALIAN RHODODENDRON SOCIETY, Attn: Mr Graeme Eaton, 1386 Mount Dandenong Tourist Road, Mount Dandenong 3767, Victoria, Australia.
Regional Representation, Japan:
THE JAPAN RHODODENDRON SOCIETY, Attn: Mr Y. Sasaki, 172–9 Nissato–machi, Souka City, Saitama Prefecture, Japan.
Regional Representation, New Zealand:
THE NEW ZEALAND RHODODENDRON ASSOCIATION, Attn: Mr M. D. Cullinane, P.O. Box 161, Awanui 05522, New Zealand.
Regional Representation, North America:
THE AMERICAN RHODODENDRON SOCIETY, Attn: Mrs J. W. Murray, 21 Squire Terrace, Colts Neck, New Jersey 07722, United States of America.
Rosa L. (1955)
THE AMERICAN ROSE SOCIETY, Attn: Mr Michael C. Kromer, American Rose Society, P.O. Box 30,000, Shreveport, Louisiana 71130–0030, United States of America. {Fax: (1) 318 938 5405}
Regional Representation, Australia:
NATIONAL ROSE SOCIETY OF AUSTRALIA, 271–B Belmore Road, North Balwyn, Victoria 3014.
Regional Representation, France:
J. Laperrière, Route Nationale 6 Chesnes, 38 Saint–Quentin–Fallavier.
Regional Representation, Germany:
Dr O. Bunemann, 4600 Dortmund 1, Am Kaiserhain 25.
Regional Representation, India:
INDIAN AGRICULTURAL RESEARCH INSTITUTE, Attn: B. Singh, Division of Horticulture, Indian Agricultural Research Institute, New Delhi 12.
Regional Representation, Italy:
ASSOCIAZIONE ITALIANA DELLA ROSA, Villa Reale, 20052 Monza.
Regional Representation, Japan:
JAPAN ROSE SOCIETY, Attn: Dr Takeo Nagata, 3–9–5 Oyamadai Setagaya–ku, Tokyo 158.
Regional Representation, The Netherlands:
RAAD VOOR HET KWEKERSRECHT, Postbus 104, NL–6700 AC Wageningen. {Fax: (31) 3174 258 67}
Regional Representation, New Zealand:
NATIONAL ROSE SOCIETY OF NEW ZEALAND, Attn: Heather MacDonnell, P.O. Box 66, 17 Erin Street, Bunnythorpe, Palmerston North.
Regional Representation, Republic of South Africa:
THE ROSE SOCIETY OF SOUTH AFRICA, Postbus 28188, 0132 Sunnyside, Pretoria.
Regional Representation, Switzerland:
CIOPORA, 4 Place Neuve, Geneva.
Regional Representation, United Kingdom:
THE ROYAL NATIONAL ROSE SOCIETY, Attn: Mrs Jill Bennell, The Royal National Rose Society, Chiswell Green, St Albans, Hertfordshire AL2 3NR {Fax: (44) 1727 850360}
Rosularia (DC.) Stapf (?1980)
THE SEMPERVIVUM SOCIETY, Attn: Mr Peter J. Mitchell, 11 Wingle Tye Road, Burgess Hill, West Sussex RH15 9HR, United Kingdom. {Fax: (44) 1444 236848}
Saintpaulia H. Wendl. (1966)
THE AFRICAN VIOLET SOCIETY OF AMERICA INC., Attn: Ms Iris Keating, 149 Loretto Court, Claremont, California 91711, United States of America.

Saxifraga L. (1994)
> THE SAXIFRAGE SOCIETY, Attn: Mr David Victor, The Old Stables, Church Lane, Hockliffe, Leighton Buzzard, Bedfordshire, LU7 9NL {Fax: (44) 1525 210070}

Sciadopitys Siebold & Zucc., see under Conifers

Scilla L., see under Bulbous plants etc.

Sempervivum L. (?1980)
> THE SEMPERVIVUM SOCIETY, Attn: Mr Peter J. Mitchell, 11 Wingle Tye Road, Burgess Hill, West Sussex RH15 9HR, United Kingdom. {Fax: (44) 1444 236848}

Sequoia Endl., see under Conifers

Sequoiadendron Buchholz, see under Conifers

Serruria Burm. ex Salisb., see under *Proteaceae*

Sinningia Nees, see under *Gesneriaceae*

Smithiantha Kuntze, see under *Gesneriaceae*

Solenophora Benth., see under *Gesneriaceae*

Sorocephalus R.Br., see under *Proteaceae*

Sparaxis Ker Gawl., see under Bulbous plants etc.

Spatalla Salisb., see under *Proteaceae*

Sprekelia Heist., see under Bulbous plants etc.

Sternbergia Waldst. & Kit., see under Bulbous plants etc.

Streptanthera Sweet (=*Sparaxis* Ker Gawl.), see under Bulbous plants etc.

Streptocalyx Beer, see under *Bromeliaceae*

Streptocarpus Lindl., see under *Gesneriaceae*

×*Streptolarium* D. Beadle (*Nidularium* Lem. × *Streptocalyx* Beer), see under *Bromeliaceae*

×*Streptomea* E. L. Sm. (*Aechmea* Ruiz & Pav. × *Streptocalyx* Beer), see under *Bromeliaceae*

Syringa L. (1958)
> ROYAL BOTANICAL GARDENS, Attn: Mr Freek Vrugtman, Royal Botanical Gardens, Box 399, Hamilton, Ontario L8N 3H8, Canada. {Fax: (1) 905 577 0375}

Tagetes L. (?1970)
> INSTITUT FÜR ZIERPLFLANZENBAU, Attn: Prof. Dr K. Zimmer and Mrs Renate Noack, Institut für Zierpflanzenbau, Herrenhäuser Strasse 2, D–30419 Hannover-Herrenhausen, Germany. {Fax: (49) 511 76 22 654}

Taiwania Hayata, see under Conifers

Taxodium Rich., see under Conifers

Taxus L., see under Conifers

Tecophilaea Bertero ex Colla, see under Bulbous plants etc.

Testudinaria Salisb. (=*Dioscorea* L.), see under Bulbous plants etc.

Tetraclinis Mast., see under Conifers

Thuja L., see under Conifers

Thujopsis Siebold & Zucc. ex Endl., see under Conifers

Tigridia Juss., see under Bulbous plants etc.

Tillandsia L., see under *Bromeliaceae*

Torreya Arn., see under Conifers

Trichantha Hook., see under *Gesneriaceae*

Trifurcia Herb. (=*Herbertia* Sweet), see under Bulbous plants etc.

Triteleia Douglas ex Lindl., see under Bulbous plants etc.

Tritonia Ker Gawl., see under Bulbous plants etc.

Tsuga Carrière, see under Conifers

Tulipa L., see under Bulbous plants etc.

Ulmus L. (1962)
> THE ARNOLD ARBORETUM, Attn: Dr Stephen A. Spongberg, Arnold Arboretum, Jamaica Plain, Massachusetts 02130, United States of America. {Fax: (1) 617 524 1418, E-mail: spongberg@arnarb.harvard.edu}

Urceolina Roxb., see under Bulbous plants etc.
Vallota Salisb. ex Herb. (=*Cyrtanthus* Aiton), see under Bulbous plants etc.
Veltheimia Gled., see under Bulbous plants etc.
Vexatorella Rourke, see under *Proteaceae*
Viburnum L. (1958)
    THE UNITED STATES NATIONAL ARBORETUM, U.S. National Arboretum, 3501 New York Avenue NE, Washington, DC 20002, United States of America. {Fax: (1) 202 245 45 75}
Vriesea Lindl., see under *Bromeliaceae*
×*Vrieslandsia* C. Chev. (*Tillandsia* L. × *Vriesea* Lindl.), see under *Bromeliaceae*
Weigela Thunb. (1966)
    THE ARNOLD ARBORETUM, Attn: Dr Stephen A. Spongberg, Arnold Arboretum, Jamaica Plain, Massachusetts 02130, United States of America. {Fax: (1) 617 524 1418, E–mail: spongberg@arnarb.harvard.edu}
Widdringtonia Endl., see under Conifers
Zantedeschia Spreng., see under Bulbous plants etc.
Zephyranthes Herb., see under Bulbous plants etc.
Zigadenus Michx., see under Bulbous plants etc.

APPENDIX III

DIRECTORY OF STATUTORY PLANT REGISTRATION AUTHORITIES

The following national authorities are known to register cultivar epithets as part of their statutory requirements.

Authorities who register epithets as part of the grant of plant breeders' rights acting under the international provisions of the "Union Internationale pour la Protection des Obtentions Végétales" ("International Union for the Protection of New Varieties of Plants" or UPOV)[1] are prefixed with ■.

The International Commission for the Nomenclature of Cultivated Plants will be glad to be informed of any other similar statutory authorites which might be included in further editions of this *Code*.

Note 1: The abbreviations shown after each country have no official status but are provided to fulfill the purposes of Recommendation 24B.1 of the Rules in this *Code*.

Note 2: The telephone and fax numbers provided are international numbers: local callers do not dial the country code shown in brackets and may have to prefix the remaining number with another digit.

■ ARGENTINA   [ARPBR]
  Instituto Naçional de Semillas, Ministerio de Economía, Secretaría de Agricultura, Ganadería y Pesca, Avda. Paseo Colón 922 – 3. Piso, 1063 Buenos Aires
  Telephone: (54) 1 362 39 88 or (54) 1 349 24 17
  Fax: (54) 1 349 24 17

■ AUSTRALIA   [AUPBR]
  The Registrar, Plant Variety Rights, P. O. Box 858, Canberra, A. C. T. 2601
  Telephone: (61) 6 272 42 28
  Fax: (61) 6 272 36 50

■ AUSTRIA   [ATBPR]
  Leiter des Sortenschutzamtes, Bundesamt und Forschungszentrum für Pflanzenbau, Alliiertenstrasse 1, Postfach 64, A–1201 Wien
  Telephone: (43) 1 211 13
  Fax: (43) 1 216 20 56

---

[1] Address: 34 chemin des Columbette, CH-1211 Geneva 20, Switzerland

BELARUS [BYNA]
  Belarus Patent Office, 66 pr. F. Skoriny, Minsk 220072
  Telephone: (7) 0172 395 053
  Fax: (7) 0172 394 130

■ BELGIUM [BEPBR]
  Ministère des classes myennes et de l'agriculture, Service de la protection des
  obtentions végétales et des catalogues nationaux, Tour WTC/3 - 6ème étage,
  Avenue Simon Bolivar 30, B–1210 Bruxelles
  Telephone: (32) 2 208 37 28 or (32) 2 208 37 22
  Fax: (32) 2 208 37 05

■ CANADA [CAPBR]
  The Commissioner of Plant Breeders' Rights, Agriculture and Agri–Food Canada,
  Plant Industry Directorate, Plant Products Division, 3rd Floor, East Court, Camelot
  Court, 59 Camelot Drive, Nepean, Ontario, K1A 0Y9
  Telephone: (1) 613 952–80 00
  Fax: (1) 613 992–52 19

CHILE [CLNA]
  Servicio Agrícola y Ganadero, Departamento de Semillas, Avda. Bulnes 140,
  Santiago
  Telephone: (56) 2 696 2996
  Fax: (56) 2 696 6480

COLOMBIA [CONA]
  Division de Semillas, Instituto Colombiano Agropecuario (ICA), Ministerio de
  Agricultura, Oficina 413, Calle 37 N. 8-43, Santa Fe de Bogotá
  Telephone: (57) 1 232 4697 or (57) 1 285 5520
  Fax: (57) 1 285 4351

■ CZECH REPUBLIC [CZPBR]
  Ministry of Economy, External Relations Department, Tesnov 17, 117 05 Prague 1
  Telephone: (42) 2 286 25 33
  Fax: (42) 2 231 44 77

■ DENMARK [DKPBR]
  Plantenyhedsnaevnet, Teglvaerksvej 10, Tystofte, DK–4230 Skaelskoer
  Telephone: (45) 53 59 61 41
  Fax: (45) 53 59 01 66

■ EUROPEAN UNION [EUPBR]
  (Temporary address) Community Plant Varieties Office, 102 Rue de la Loi, First
  Floor, B–1049 Brussels
  Telephone: (32) 2 299 19 44
  Fax: (32) 2 299 19 46

■ FINLAND [FIPBR]
  Plant Variety Board, Plant Variety Rights Office, Box 232, SF–00171 Helsinki
  Telephone (358) 0 160 3316
  Fax: (358) 0 160 2443

- FRANCE [FRPBR]
  Comité de la protection des obtentions végétales, 11 rue Jean Nicot, F–75007 Paris
  Telephone: (331) 42 75 93 14
  Fax: (331) 42 75 94 25

- GERMANY [DEPBR]
  Bundessortenamt, Postfach 61 04 40, D–30627 Hannover
  Telephone: (49) 511 95 66–5
  Fax: (49) 511 56 33 62

- HUNGARY [HUPBR]
  Office national des inventions, Országos Találmányi Hivatal, Garibaldi–u.2 – B. P. 552, H–1370 Budapest 5
  Telephone: (36) 1 111 28 93
  Fax: (36) 1 131 25 96

- IRELAND [IEPBR]
  Controller of Plant Breeders' Rights, National Variety Testing Centre, Backweston, Leixlip, Co. Kildare
  Telephone: (353) 628 06 08
  Fax: (353) 628 06 34

- ISRAEL [ILPBR]
  Plant Breeders' Rights Council, The Volcani Center, P. O. Box 6, Bet–Dagan 50 250
  Telephone: (972) 3 968 34 92
  Fax: (972) 3 968 34 92

- ITALY [ITPRP]
  Ufficio Italiano Brevetti e Marchi, Ministero dell'Industria del Commercio e dell'Artigianato, 19 via Molise, I–00187 Roma
  Telephone: (39) 6 47 05 1
  Fax: (39) 6 47 05 30 35

- JAPAN [JPPBR]
  Director of Seeds and Seedlings Division, Agricultural Production Bureau, Ministry of Agriculture Forestry and Fisheries, 1–2–1 Kasumigaseki – Chiyoda–ku Tokyo
  Telephone: (81) 3 591 95 24
  Fax: (81) 3 502 65 72

KENYA [KENA]
  Ministry of Agriculture, P. O. Box 30028, Nairobi
  Telephone: Unknown
  Fax: Unknown

- NETHERLANDS [NLPBR]
  Raad voor het Kwekersrecht, Postbus 104, NL–6700 AC Wageningen
  Telephone: (31) 3174 190 31
  Fax: (31) 3174 258 67

- NEW ZEALAND [NZPBR]
  Commissioner of Plant Variety Rights, Plant Variety Rights Office, P. O. Box 24, Lincoln
  Telephone: (64) 3 325 6355
  Fax: (64) 3 325 2946

■ NORWAY [NOPBR]
  Plantesortsnemnda, Fellesbygget, N-1432 As
  Telephone: (47) 64 94 75 04
  Fax: (47) 64 94 02 08

PARAGUAY [PYNA]
  Dirección de Semillas (DISA)–MAG, Gaspar R. De Francia No. 685, c/ Mcal.
  Estigarribia, San Lorenzo
  Telephone: (595) 582 2201
  Fax: (595) 584 645

■ POLAND [PLPBR]
  The Director, Research Center of Cultivars Testing (COBORU), 63–022 Słupia
  Wielka
  Telephone: (48) 665 535 58 or (48) 665 523 41
  Fax: (48) 665 535 58

PORTUGAL [PTNA]
  Centro Nacional de Registo de Variedades Protegidas (CENARVE), Tapada da
  Ajuda, Edificio II, 1300 Lisboa
  Telephone: (351) 362 16 07
  Fax: (351) 362 16 06

RUSSIAN FEDERATION [RUNA]
  State Commission of the Russian Federation for Selection Achievements Test and
  Protection, 3a Orlikov per., 107139 Moscow
  Telephone: (7) 095 204 49 26 or (7) 095 204 48 53
  Fax: (7) 095 207 86 26

■ SLOVAKIA [SKPBR]
  Ministry of Agriculture, Dobrovicova 12, 812 66 Bratislava
  Telephone: (42) 7 32 50 52
  Fax: (42) 7 32 50 52

■ SOUTH AFRICA [SAPBR]
  Department of Agriculture, Directorate of Plant Quality Control, Private Bag X179,
  Pretoria 0001
  Telephone: (27) 12 319 60 00
  Fax: (27) 12 319 60 55

■ SPAIN [ESPBR]
  Registro de Variedades, Instituto Naçional de Semillas y Plantas de Vivero, José
  Abascal 56, E–28003 Madrid
  Telephone: (34) 1 347 69 00
  Fax: (34) 1 442 82 64

■ SWEDEN [SEPBR]
  Statens växtsortnämnd, Box 1247, S–171 24 Solna
  Telephone: (46) 8 730 66 30 or (46) 8 730 66 40
  Fax: (46) 8 833 170

■ <u>SWITZERLAND</u> [CHPBR]
　Bundesamt für Landwirtschaft, Büro für Sortenschutz, Mattenhofstr. 5, CH–3003
　Bern
　Telephone: (41) 31 322 25 24
　Fax: (41) 31 322 26 34

<u>UKRAINE</u> [UANA]
　The State Commission of Ukraine for Sort Testing and Protection of Plants,
　Suvorov St. 9, 252010 Kiev
　Telephone: (7) 044 290 31 91
　Fax: (7) 044 290 42 62 or (7) 044 290 51 34

■ <u>UNITED KINGDOM</u> [UKPBR]
　The Plant Variety Rights Office, White House Lane, Huntingdon Road, Cambridge
　CB3 0LF
　Telephone: (44) 1223 27 71 51
　Fax: (44) 1223 34 23 86

<u>UNITED STATES OF AMERICA</u> [USPP]
　The Commissioner of Patents and Trademarks, Patent and Trademark Office, Box
　4, Washington D.C. 20231
　Telephone: (1703) 305 93 00
　Fax: (1703) 305 88 85

■ <u>UNITED STATES OF AMERICA</u> [USPBR]
　The Commissioner, Plant Variety Protection Service, Agricultural Marketing
　Service, Department of Agriculture, Beltsville, Maryland 20705–2351
　Telephone: (1) 301 504 55 18
　Fax: (1) 301 504 52 91

■ <u>URUGUAY</u> [UYPBR]
　Ministerio de Ganadería Agricultura y Pesca, Dirección General – Servicios
　Agrícolas, Unidad de Semillas, Avda. Millán 4703, 12.900 Montevideo
　Telephone: (598) 2 39 79 24
　Fax: (598) 2 39 78 32

<u>ZIMBABWE</u> [ZWNA]
　Seed Services, Department of Research and Specialist Services, Box 8100, Cause-
　way, Harare
　Telephone: (263) 4 720 370 or (263) 4 791 223
　Fax: (263) 4 728 317

APPENDIX IV

LIST OF DENOMINATION CLASSES

The denomination class as defined in this *Code* is the genus or nothogenus unless listed below.

†: those originally designated by UPOV, the 'Union Internationale pour la Protection des Obtentions Végétales' ('International Union for the Protection of New Varieties of Plants' or UPOV), and accepted by The International Commission for the Nomenclature of Cultivated Plants in this *Code*.

IRA: those denomination classes covered by an International Registration Authority (see Appendix II for a directory of IRAs).

Anacardiaceae Lindl.
  Class: *Mangifera indica* L. **IRA**
  Class: *Mangifera* L.: other than *Mangifera indica* L.

Apiaceae Lindl. (nom. alt.: *Umbelliferae* Juss.)
†Class: *Anethum* L., *Carum* L., *Foeniculum* Mill.
†Class: *Anthriscus* Pers., *Petroselinum* Hill
†Class: *Daucus* L., *Pastinaca* L.

Asteraceae Dumort. (nom. alt.: *Compositae* Giseke)
†Class: *Cichorium* L., *Lactuca* L., *Valerianella* Mill.
†Class: *Helianthus annuus* L.
†Class: *Helianthus tuberosus* L.
†Class: *Helianthus* L.: other than *H. annuus* L. or *H. tuberosus* L.

Brassicaceae Burnett (nom. alt.: *Cruciferae* Juss.)
†Class: *Brassica campestris* L., *B. juncea* (L.) Czern. & Coss., *B. napus* L., *B. nigra* (L.) W. D. J. Koch, *B. rapa* L., *Sinapis* L.
†Class: *Brassica oleracea* L.
†Class: *Brassica* L.: other than *B. campestris* L., *B. juncea* (L.) Czern. & Coss., *B. napus* L., *B. nigra* (L.) W. D. J. Koch, *B. rapa* L. or *B. oleracea* L.

Cactaceae Juss.
†Class: *Epiphyllum* Haw., *Rhipsalidopsis* Britton & Rose, *Schlumbergera* Lem., *Zygocactus* Schum.

Chenopodiaceae Vent.
†Class: *Beta vulgaris* L. Fodder Beet Group, *Beta vulgaris* L. Sugar Beet Group
†Class: *Beta vulgaris* L. Spinach Beet Group, *Beta vulgaris* L. Beetroot Group, *Beta vulgaris* L. Mangold Group
†Class: *Beta* L.: other than *B. vulgaris* L. Fodder Beet Group, *B. vulgaris* L. Sugar Beet Group or *B. vulgaris* L. Spinach Beet Group, *B. vulgaris* L. Beetroot Group, *B. vulgaris* L. Mangold Group

Compositae Giseke: see *Asteraceae* Lindl.

Crassulaceae DC.
  Class: *Jovibarba* Opiz, *Rosularia* (DC) Stapf, *Sempervivum* L. **IRA**

Cruciferae Juss.: see *Brassicaceae* Burnett

Cucurbitaceae Juss.
  †Class: *Citrullus* Schrad., *Cucumis melo* L., *Cucurbita* L.
  †Class: *Cucumis sativus* L.
  †Class: *Cucumis* L.: other than *C. melo* L. or *C. sativus* L.

Cupressaceae Rich. ex Bartl.
  Class: *Chamaecyparis* Spach, *Cupressus* L., ×*Cupressocyparis* Dallim. **IRA**

Ericaceae Juss.
  Class: *Andromeda* L., *Bruckenthalia* Rchb., *Calluna* Salisb., *Daboecia* D. Don,
           Erica L. **IRA**
  Class: *Azalea* L., *Rhododendron* L. **IRA**

Fabaceae Lindl. (nom. alt.: *Leguminosae* Juss.)
  †Class: *Lotus* L., *Medicago* L., *Onobrychis* Mill., *Ornithopus* L., *Trifolium* L.
  †Class: *Lupinus albus* L., *L. angustifolius* L., *L. luteus* L.
  †Class: *Lupinus* L.: other than *Lupinus albus* L., *L. angustifolius* L., *L. luteus* L.
  †Class: *Vicia faba* L.
  †Class: *Vicia* L.: other than *V. faba* L.

Gramineae Juss.: see *Poaceae* (R. Br.) Barnhart

Iridaceae Juss.
  Class: *Gladiolus* L.: large-flowering types **IRA**
  Class: *Gladiolus* L.: species & early-flowering types **IRA**
  Class: *Iris* L.: bulbous types **IRA**
  Class: *Iris* L.: non-bulbous types **IRA**

Leguminosae Juss.: see *Fabaceae* Lindl.

Malvaceae Small ex Britton & Small
  Class: *Hibiscus rosa-sinensis* L. **IRA**
  Class: *Hibiscus* L.: other than *Hibiscus rosa-sinensis* L.

Poaceae (R. Br.) Barnhart (nom. alt.: *Gramineae* Juss.)
  †Class: *Avena* L., *Hordeum* L., *Secale* L., ×*Triticale* Müntzing, *Triticum* L.
  †Class: *Panicum* L., *Setaria* P. Beauv.
  †Class: *Sorghum* Moench, *Zea* L.
  †Class: *Agrostis* L., *Alopecurus* L., *Arrhenatherum* P. Beauv., *Bromus* L.,
           Cynosurus L., *Dactylis* L., *Festuca* L., *Lolium* L., *Phalaris* L., *Phleum* L.
           Poa L., *Trisetum* Pers.

Ranunculaceae Juss.
  Class: *Delphinium* L.: perennials **IRA**
  Class: *Delphinium* L.: non-perennials

Rosaceae Juss.
Class: *Malus domestica* Borkh.
Class: *Malus* Mill.: other than *Malus domestica* Borkh. **IRA**
Class: *Potentilla fruticosa* L. *sensu lato* **IRA**
Class: *Potentilla* L.: other than *P. fruticosa* L. *sensu lato*

Solanaceae Juss.
†Class: *Nicotiana rustica* L. & *N. tabacum* L.
†Class: *Nicotiana* L.: other than *N. rustica* L. & *N. tabacum* L.
†Class: *Solanum tuberosum* L.
†Class: *Solanum* L.: other than *S. tuberosum* L.

Umbelliferae Juss.: see **Apiaceae** Lindl.

NOTE: In addition to the denomination classes listed above, UPOV has additional denomination classes in which the duplication of cultivar names is not permitted for the purposes of variety denomination for Plant Breeders' Rights. These are not currently accepted by The International Commission for the Nomenclature of Cultivated Plants in this *Code*.

Orchidaceae Juss.
 Class: The family
Pinaceae Lindl.:
 Class: *Abies* Mill., *Larix* Mill., *Picea* A. Dietr., *Pinus* L., *Pseudotsuga* Carrière
Proteaceae Juss.
 Class: The family

APPENDIX V

LIST OF CONSERVED EPITHETS

Epithets are permanently conserved by action of the International Commission for the Nomenclature of Cultivated Plants upon written request from an International Registration Authority or any other interested party who makes a submission to the Commission.

Submissions, which must contain written details for each case to be considered, must be made to the Secretary to the Commission: Drs W. L. A. Hetterscheid, Vaste Keurings Commissie, Linnaeuslaan 2a, 1431 JV Aalsmeer, The Netherlands. {Fax: (31) 297 352 205}

The following list remains permanently open for additions.

VIBURNUM L. (*Caprifoliaceae*)

'Sterile' in *V. macrocephalum* 'Sterile'
syn: *V. macrocephalum* Fortune var. *macrocephalum*
syn: *V. macrocephalum* var. *sterile* Dippel, Handb. Laubh. 1: 178. 1889 *nom. illegit.*
syn: *V. keteleeri macrocephalum* Carrière, Rev. Hort. 1863: 271. 1863
syn: *V. fortunei* hort. ex Nicholson, Ill. Dict. Gard. 4: 155. 1887 *pro syn.*

APPENDIX VI

HERBARIA MAINTAINING STANDARDS

The following herbaria are known to hold collections of Standards as defined in this *Code*.

This initial list is known to be far from complete: the editor of this *Code* will be pleased to receive details of further herbaria who wish to be added to this list.

Note 1: This Appendix does not cover those institutions holding germplasm Standards: a list of such institutions may be found in Bettencourt, E., Hazekamp, Th., Perry, M. C., (eds), *Directory of germplasm collections* (several vols), International Board for Plant Genetic Resources (IBPGR), Rome, Italy.

Note 2: The abbreviations given before each herbarium name are those in Holmgren, P. K., Holmgren, N. H., Barnett, L. C. (eds), *Index herbariorum* ed. 8. 1990 (*Regnum vegetabile 120*) New York Botanical Garden, U. S. A.

AUSTRALIA

(AD) Botanic Gardens and State Herbarium of South Australia, North Terrace, Adelaide, South Australia {Fax: (61) 8 223 1809}
Contact: Trevor Christensen, Horticultural Botanist

(CANB) Australian National Herbarium, Centre for Plant Biodiversity Research, Division of Plant Industry, CSIRP, GPO Box 1600, Canberra, ACT 2601 {Fax: (61) 6 246 5249}
Contact: Lyn A. Craven {E-mail: craven@pi.csiro.au}

(MEL) Royal Botanic Gardens Melbourne, Birdwood Road, South Yarra, Victoria 3141 {Fax: (61) 3 9252 2350}
Contact: Roger Spencer {E-mail: U6067804@ucsvc.ucs.unimelb.edu.au}

(NSW) National Herbarium of New South Wales, Royal Botanic Gardens Sydney, Mrs Macquaries Road, Sydney, New South Wales 2000 {Fax: (61) 2 251 7231}
Contact: Mrs Gwen Harden, Curator–Manager of the Herbarium {E-mail: gwen@rbgsyd.gov.au}

CANADA

(DAO) Agriculture & Agri–Food Canada, Centre for Land and Biological Resources Research, William Saunders Building, Ottowa, Ontario K1A 0C6 {Fax: (1) 613 759 1599}
Contact: Gisele Mitrow, Collections Manager {E-mail: mitrog@ncccot.agr.ca}

(HAM) Royal Botanical Gardens, PO Box 399, Hamilton, Ontario L8N 3H8 {Fax: (1) 905 577 0375}
Contact: Freek Vrugtman

THE NETHERLANDS

(L) Rijksherbarium/Hortus Botanicus, PO Box 9514, 2300 RA Leiden {Fax: (31) 3171 5273511}
Contact: Professor Dr P. Baas {E–mail: baas@rulrhb.leidenuniv.nl}

(WAG) Herbarium Vadense/Department of Plant Taxonomy, Agricultural University, PO Box 8010, 6700 ED Wageningen {Fax: (31) 3174 84917}
Contact: Folkert Aleva, Herbarium Manager {E–mail: folkert.aleva@ algem.pt.wau.nl or vadense@algem.pt.wau.nl}

NEW ZEALAND

(CHR) The Herbarium, Manaaki Whenua – Landcare Research, Gerald Street, PO Box 69, Lincoln 8152 {Fax: (64) 3 325 2418}
Contact: Dr Murray J. Parsons, Herbarium Keeper {E–mail: parsonsm@landcare.cri.nz}

NORWAY

(BG) Herbarium, Botanical Institute, University of Bergen, Allégaten 41, N–5007 Bergen {Fax: (47) 55 31 2238}
Contact: Prof. Dr P. M. Jørgensen

THE PHILIPPINES

(CSCS) The Herbarium, Cebu State College of Science and Technology, College of Agriculture, Lahug, Cebu City, 6000 Cebu {Fax: None}
Contact: Inocencio E. Buot, Jr
Or: Rebecca DC. Manalastas, Officer–in–Charge

SLOVAK REPUBLIC

(SAV) Institute of Botany, Slovak Academy of Sciences, Herbarium Dúbravská cesta 14, SK–842 23 Bratislava {Fax: (42) 7 371948}
Contact: Dr Karol Marhold, Keeper of the Herbarium {E–mail: botukmar@savba.sav-ba.sk}

SWITZERLAND

(G) Herbarium, Conservatoire et Jardin botaniques de la Ville de Genève, Case postale 60, CH–1292 Chambésy/Genève {Fax: (41) 22 738 45 97}
Contact: A. Charpin, Head Curator of Herbarium

UNITED KINGDOM

(CGG) Garden Herbarium, University Botanic Garden, Cory Lodge, Bateman Street, Cambridge CB2 1JF {Fax: 01223 336278}

(E) The Royal Botanic Garden Edinburgh, Inverleith Row, Edinburgh EH3 5LR {Fax: (44) 131 552 0382}
Contact: Dr Crinan Alexander, Horticultural Taxonomist {E–mail: crinan@rbge.org.uk}

The Sir Harold Hillier Gardens and Arboretum, Jermyns Lane, Ampfield, Romsey, Hampshire SO51 0QA {Fax: (44) 1794 368027}
Contact: Mr Allen Coombes, Botanist to the Gardens and Arboretum

(K) Royal Botanic Gardens, Kew, Richmond, Surrey TW9 3AB {Fax: (44) 181 948 1197}
Contact: Susyn Andrews, Horticultural Taxonomist

(LIV) Herbarium, Botany Department, Liverpool Museum, William Brown Street, Liverpool L3 8EN {Fax: (44) 151 478 4390}
Contact: John. R. Edmondson {E–mail: 100145.554@compuserve.com}

(RDG) School of Plant Sciences, University of Reading, Whiteknights, PO Box 221, Reading, Berkshire RG6 6AS {Fax: (44) 1734 753676}
Contact: Dr Stephen L. Jury, Herbarium Curator

(WSY) The Royal Horticultural Society's Garden, Wisley, Woking, Surrey GU23 6QB {Fax: (44) 1483 211750}
Contact: Mrs Diana Miller, Keeper of the Herbarium

UNITED STATES OF AMERICA

(BH) Herbarium, L. H. Bailey Hortorium, 462 Mann Library, Cornell University, Ithaca, New York 14853–4301 {Fax: (1) 607 255 7979}
Contact: Edward A. Cope {E–mail:eac5@cornell.edu}

(BISH) Herbarium Pacificum, Bishop Museum, Department of Natural Sciences/Botany, 1525 Bernice Street, PO Box 19000, Honolulu, Hawai'i, 96817–0916 {Fax: (1) 808 841 8968}
Contact: Walter Appleby, Collections Manager {E–mail: botany@bishop.bishop.hawaii-.org}

(BKL) Brooklyn Botanic Garden, 1000 Washington Avenue, Brooklyn, New York 11225–1099 {Fax: (1) 718 941 4044 ext. 289}
Contact: Kerry Barringer, Curator of the Herbarium {E–mail: kerryb@panix.com}

(CM) Section of Botany, Carnegie Museum of Natural History, 4400 Forbes Avenue, Pittsburgh, Pennsylvania 15213 {Fax: (1) 412 622 8837}
Contact: Sue Thompson, Collection Manager {E–mail: thompsons@clpgh.org}

(CONN) George Safford Torrey Herbarium, Ecology and Evolutionary Biology, U–43, 75 North Eagleville Road, University of Connecticut, Storrs, Connecticut 06269–3043 {Fax: (1) 203 486 6364}
Contact: Gregory J. Anderson {E–mail %: decarli@uconnvm.uconn.edu}

(DES) Desert Botanical Garden of Arizona, 1201 North Galvin Parkway, Phoenix, Arizona 85008 {Fax: (1) 602 481 8124}
Contact: Wendy Hodgson, Herbarium Curator {E-mail: adjbc@asuvm.inre.asu.edu}

(DOV) Claude E. Phillips Herbarium, Department of Agriculture & Natural Resources, Delaware State University, Dover, Delaware 19901-2277 {Fax: (1) 302 739 4997}
Contact: Dr Arthur O. Tucker {E-mail: aotucker@igc.apc.org}

(MARY) Norton-Brown Herbarium, Department of Plant Biology, University of Maryland, College Park, Maryland 20742-5815 {Fax: (1) 301 314 9082}
Contact: James L. Reveal {E-mail: jr@umail.umd.edu}

(MO) Herbarium, Missouri Botanical Garden, PO Box 299, Saint Louis, Missouri 63166-0299 {Fax: (1) 314 577 9596}
Contact: James C. Solomon, Curator of the Herbarium {E-mail: solomon@mobot.org}
Or:
Alan W. Lievens, Horticultural Taxonomist {E-mail: alievens@ridgway.mobot.org}

(MOR) The Morton Arboretum, Lisle, Illinois 60532 {Fax: (1) 708 719 2433}
Contact: William J. Hess {E-mail: morherb@cedar.cic.net}

(MU) Willard Sherman Turrell Herbarium, Department of Botany, Miami University, Oxford, Ohio 45045 {Fax: (1) 513 529 4243}
Contact: Dr Michael A. Vincent, Curator {E-mail: vincenma@muohio.edu}

(NA) United States National Arboretum, 3501 New York Avenue North East, Washington, DC 20002 {Fax: (1) 202 245 45 75}
Contact: (position vacant at time of publishing)

(OSC) The Herbarium, Department of Botany and Plant Pathology, Cordley Hall 2082, Oregon State University, Corvallis, Oregon 97331-2902 {Fax: (1) 503 737 3573}
Contact: Dr Richard Halse, Herbarium Curator {E-mail: halser@bcc.orst.edu}

(PH) Academy of Natural Sciences of Philadelphia, 1900 Benjamin Franklin Parkway, Philadelphia, Pennsylvania 19103 {Fax: (1) 215 299 1028}
Contact: Alfred Ernest Schuyler {E-mail: schuyler@say.acnatsci.org}

(RSA-POM) Rancho Santa Ana Botanic Garden, 1500 North College Avenue, Claremont, California 91711 {Fax: (1) 909 626 7670}
Contact: Steve Boyd, Administrative Curator of the Herbarium {E-mail: boyds@cgs.-edu}
Or: Bart O'Brien, Director of Horticulture, Phone: (1) 909 625 8767

(WTU) Hyde Hortorium, Center for Urban Horticulture, GF-15 University of Washington, Seattle, Washington 98195 {Fax: (1) 206 685 2692}
Contact: Clement W. Hamilton, Director {E-mail: cwh@u.washington.edu}

APPENDIX VII

THE NOMENCLATURAL FILTER

In this *Code*, unless otherwise indicated, the words "epithet" or "name" mean an epithet or name that has been established, whether it is accepted or not. In the following filters the words epithet and name appears in quotation marks until the point of establishment is realized. Only then may one consider whether an epithet or name is accepted or not.

By following the path below, one may check that an "epithet" or "name" is (a) established and (b) accepted.

1. CULTIVAR EPITHETS

1   Is the "epithet" denominated under Plant Breeders' Rights?
    YES: ⇒ **35**   No: ⇒ **2**

2   Is the "epithet" given under an award of Plant Patent?
    YES: ⇒ **35**   No: ⇒ **3**

3   Is the "epithet" actually a trade-mark?
    No: ⇒ **4**   YES: **STOP**   COMMENT: TRADE-MARKS ARE NEVER TO BE CONSIDERED AS CULTIVAR EPITHETS.

4   Is the "epithet" another sort of trade designation? (Art. 11.1, 28.1, Note 1)
    No: ⇒ **5**   YES: **STOP**   COMMENT: TRADE DESIGNATIONS ARE MARKETING DEVICES USED IN PLACE OF THE ACCEPTED EPITHET AND ARE NOT CULTIVAR EPITHETS.

5   Has the "epithet" been conserved by the International Commission for the Nomenclature of Cultivated Plants? (Art. 14.1)
    YES: ⇒ **35**   No: ⇒ **6**

6   Has the "epithet" been used before in the denomination class even as a trade designation? (Art. 17.2)
    YES: ⇒ **7**   No: ⇒ **8**

7   Has use of the "epithet" been sanctioned by an International Registration Authority? (Art. 14.3-14.4, 26.2)
    YES: ⇒ **35**   No: **STOP**   COMMENT: EPITHETS MAY NOT BE USED MORE THAN ONCE IN A DENOMINATION CLASS UNLESS SANCTIONED OR CONSERVED.

8   Has the "epithet" appeared in a publication that is printed or is similarly
duplicated matter? (Art. 21.1-21.3)
Y<small>ES</small>: ⇒ **9**   N<small>O</small>: **STOP**   C<small>OMMENT</small>: R<small>EJECT AS NOT PUBLISHED</small>.

9   Was the "epithet" published on or after the starting point for the group concerned? (Art. 13.1)
Y<small>ES</small>: ⇒ **10**   N<small>O</small>: **STOP**   C<small>OMMENT</small>: F<small>IND A LATER PLACE OF PUBLICATION</small>.

10   After 1958, was the publication in which the "epithet" appeared dated at least
to the year? (Art. 23.1)
Y<small>ES</small>: ⇒ **11**   N<small>O</small>: **STOP**   C<small>OMMENT</small>: R<small>EJECT AS NOT ESTABLISHED</small>
P<small>UBLICATIONS SINCE</small> 1958 <small>MUST BE DATED</small> - <small>FIND A</small>
L<small>ATER PUBLICATION THAT IS DATED</small>.

11   After 1958, is the "epithet" in a modern language? (Art. 17.9)
Y<small>ES</small>: ⇒ **13**   N<small>O</small>: ⇒ **12**

12   Is the "epithet" taken from a Latin epithet at the rank of species or below
which is established (validly published) and acceptable in conformity with the
ICBN for a taxon subsequently reclassified as a cultivar? (Art. 17.3)
Y<small>ES</small>: ⇒ **13**   N<small>O</small>: **STOP**   C<small>OMMENT</small>: R<small>EJECT AS NOT ESTABLISHED</small>.

13   After 1958 and before 1995, is the "epithet" of more than three words? (Art.
17, Note 3)
N<small>O</small>: ⇒ **14**   Y<small>ES</small>: **STOP**   C<small>OMMENT</small>: R<small>EJECT AS NOT ESTABLISHED</small>.

14   After 1995, are there more than 10 syllables or more than 30 characters in the
"epithet"? (Art. 17.10)
N<small>O</small>: ⇒ **15**   Y<small>ES</small>: **STOP**   C<small>OMMENT</small>: R<small>EJECT AS NOT ESTABLISHED</small>.

15   Is the "epithet" only made up of common descriptive words? (Art. 17.11)
N<small>O</small>: ⇒ **16**   Y<small>ES</small>: **STOP**   C<small>OMMENT</small>: R<small>EJECT AS NOT ESTABLISHED</small>.

16   After 1995, is the "epithet" very similar to or confusable with another in the
same genus or denomination class? (Art 17.12)
N<small>O</small>: ⇒ **17**   Y<small>ES</small>: **STOP**   C<small>OMMENT</small>: R<small>EJECT AS NOT ESTABLISHED</small>.

17   After 1958, is the "epithet" the same as the botanical, common or vernacular
name of any genus or nothogenus? (Art. 17.13)
N<small>O</small>: ⇒ **18**   Y<small>ES</small>: **STOP**   C<small>OMMENT</small>: R<small>EJECT AS NOT ESTABLISHED</small>.

18   After 1958, is the "epithet" a common or vernacular name of a species within
its genus or nothogenus? (Art. 17.13)
N<small>O</small>: ⇒ **19**   Y<small>ES</small>: **STOP**   C<small>OMMENT</small>: R<small>EJECT AS NOT ESTABLISHED</small>.

19   After 1995, is the final word of the "epithet" the same as the botanical,
common or vernacular name of any genus or nothogenus? (Art. 17.13)
N<small>O</small>: ⇒ **20**   Y<small>ES</small>: **STOP**   C<small>OMMENT</small>: R<small>EJECT AS NOT ESTABLISHED</small>.

20   After 1995, does the "epithet" contain the botanical, common or vernacular
name of its genus or nothogenus? (Art. 17.13)
N<small>O</small>: ⇒ **21**   Y<small>ES</small>: **STOP**   C<small>OMMENT</small>: R<small>EJECT AS NOT ESTABLISHED</small>.

21 After 1995, does the "epithet" contain the common or vernacular name of a species within its genus or nothogenus? (Art. 17.13)
Yes: ⇒ **22**   No: ⇒ **23**

22 Is the included common or vernacular name of the species a permitted transliteration from Japanese? (Art. 17.14)
Yes: ⇒ **23**   No: **STOP**   COMMENT: REJECT AS NOT ESTABLISHED.

23 After 1958, do the words "variety" or "form" appear in the "epithet"? (Art. 17.15)
No: ⇒ **24**   Yes: **STOP**   COMMENT: REJECT AS NOT ESTABLISHED.

24 After 1958 does the word "var" as an abbreviation for "variegated" appear in the "epithet"? (Art. 17.15)
No: ⇒ **25**   Yes: **STOP**   COMMENT: ALTER THE WORD "VAR" TO "VARIEGATED" AND ⇒ **25**

25 After 1995, do any of the words "cross", "hybrid", "grex", "group", "maintenance", "mutant", "seedling", "selection", "sport", "strain", or the plural form of these words in any language or the words "improved" or "transformed" appear in the "epithet"? (Art. 17.16-17.17)
No: ⇒ **26**   Yes: **STOP**   COMMENT: REJECT AS NOT ESTABLISHED.

26 After 1995, is the "epithet" likely to exaggerate the merits of the cultivar? (Art. 17.18)
No: ⇒ **27**   Yes: **STOP**   COMMENT: REJECT AS NOT ESTABLISHED.

27 After 1995, does the "epithet" contain inadmissable punctuation marks? (Art. 17.19)
No: ⇒ **28**   Yes: **STOP**   COMMENT: REJECT AS NOT ESTABLISHED.

28 After 1958, is there a description or diagnosis accompanying publication of the "epithet"? (Art. 22.1)
Yes: ⇒ **30**   No: ⇒ **29**

29 Is there a reference to a previous publication of a description or diagnosis? (Art. 22.1)
Yes: ⇒ **30**   No: **STOP**   COMMENT: REJECT AS NOT ESTABLISHED.

30 Has the "epithet" definitely been accepted by the author who publishes it? (Art. 22.3)
Yes: ⇒ **31**   No: **STOP**   COMMENT: REJECT AS NOT ESTABLISHED.

31 Did the cultivar for which the "epithet" was proposed actually exist at the time of publication? (Art. 22.5)
Yes: ⇒ **32**   No: **STOP**   COMMENT: REJECT AS NOT ESTABLISHED.

32 Is the "epithet" against the wish of the originator? (Art. 22.6)
No: ⇒ **34**   Yes: ⇒ **33**

33 Did the originator knowingly distribute the cultivar without a name? (Art. 22.6)
Y ES: ⇒ **34** No: **STOP**   COMMENT: REJECT AS NOT ESTABLISHED AND USE
THE EPITHET PROPOSED BY THE AUTHOR.

34 Was another "epithet" proposed by the same author for the same cultivar simultaneously? (Art. 22.7)
No: ⇒ **35**   Y ES: **STOP**   COMMENT: REJECT AS NOT ESTABLISHED.

YOU HAVE AN ESTABLISHED CULTIVAR EPITHET

35 Has a statutory plant registration authority denominated a different epithet for the cultivar?
No: ⇒ **36** Y ES: **STOP**   COMMENT: REJECT AS NOT ACCEPTABLE; THE
STATUTORY DENOMINATION IS AUTOMATICALLY THE
ACCEPTED EPITHET.

36 Is this the earliest epithet available for the cultivar?
Y ES: ⇒ **37** No: **STOP**   COMMENT: REJECT AS NOT ACCEPTABLE AND USE THE
EARLIEST EPITHET.

37   ### YOU HAVE THE ACCEPTED CULTIVAR EPITHET

2. CULTIVAR-GROUP EPITHETS

1 Is the "epithet" actually a trade-mark?
No: ⇒ **2**   Y ES: **STOP**   COMMENT: TRADE-MARKS ARE NOT CULTIVAR-GROUP
EPITHETS.

2 Has the "epithet" appeared in a publication that is printed or similarly duplicated matter? (Art. 21.1-21.3)
Y ES: ⇒ **3**   No: **STOP**   COMMENT: EPITHETS MUST BE PROPERLY PUBLISHED.

3 Was the "epithet" published on or after the starting point for the group concerned? (Art. 13.1)
Y ES: ⇒ **4**   No: **STOP**   COMMENT: FIND A LATER PLACE OF PUBLICATION.

4 After 1958, was the publication dated at least to the year? (Art. 23.1)
Y ES: ⇒ **5**   No: **STOP**   COMMENT: REJECT AS NOT ESTABLISHED
PUBLICATIONS SINCE 1958 MUST BE DATED - FIND A
LATER PUBLICATION THAT IS DATED.

5 Is the "epithet", apart from the word "Group" (or its equivalent), based upon a established cultivar epithet? (Art. 19.6)
Y ES: ⇒ **18**   No: ⇒ **6**

6   Does the "epithet", apart from the word "Group" (or its equivalent), contain more than three words? (Art. 19.6)
     No: ⇒ 7   YES: **STOP**   COMMENT: REJECT AS NOT ESTABLISHED.

7   Is the "epithet" (apart from the word "Group") only made up of common descriptive words? (Art. 19.7)
     No: ⇒ 9   YES: ⇒ 8

8   Do the common descriptive words qualify one or more attributes of the cultivar-group? (Art. 19.7)
     YES: ⇒ 9   No: **STOP**   COMMENT: REJECT AS NOT ESTABLISHED.

9   Is the "epithet" very similar to another? (Art 17.8)
     No: ⇒ 10   YES: **STOP**   COMMENT: REJECT AS NOT ESTABLISHED.

10   Does the "epithet" (excluding the word "Group") contain the name of its genus or nothogenus or its common or vernacular equivalent? (Art. 19.9)
     No: ⇒ 11   YES: **STOP**   COMMENT: REJECT AS NOT ESTABLISHED.

11   Do any of the words "cross", "form", "grex", "group" (except as the final word), "hybrid", "mutant", "seedling", "selection", "sport", "strain", "variety", the plural form of these words, or the words "improved" and "transformed" appear in the epithet? (Art. 19.10)
     No: ⇒ 12   YES: **STOP**   COMMENT: REJECT AS NOT ESTABLISHED.

12   Is the "epithet" likely to exaggerate the merits of the cultivars in the cultivar-group? (Art. 19.11)
     No: ⇒ 13   YES: **STOP**   COMMENT: REJECT AS NOT ESTABLISHED.

13   Does the "epithet" contain inadmissible punctuation marks? (Art. 19.12)
     No: ⇒ 14   YES: **STOP**   COMMENT: REJECT AS NOT ESTABLISHED.

14   After 1958, is there a description or diagnosis accompanying the "epithet"? (Art. 22.1)
     YES: ⇒ 16   No: ⇒ 15

15   Is there a reference to a previous publication of a description or diagnosis? (Art. 22.1)
     YES: ⇒ 16   No: **STOP**   COMMENT: REJECT AS NOT ESTABLISHED.

YOU HAVE AN ESTABLISHED CULTIVAR-GROUP EPITHET

16   Has the "epithet" been used before in the denomination class? (Art. 19.2)
     No: ⇒ 17   YES: **STOP**   COMMENT: REJECT AS NOT ACCEPTABLE.

17   Does the epithet replace another with current established use? (Art. 15.2)
     No: ⇒ 18   YES: **STOP**   COMMENT: REJECT AS NOT ACCEPTABLE AND USE THE EPITHET WITH ESTABLISHED USE.

18   **YOU HAVE THE ACCEPTED CULTIVAR-GROUP EPITHET**

3. GRAFT-CHIMAERAS

1   Is the graft-chimaera between members of two or more genera?
    YES: ⇒ 2   No: ⇒ 1 IN FILTER FOR CULTIVAR EPITHETS

2   Has the "name" appeared in a publication that is printed or similarly duplicated matter? (Art. 21.1-21.3)
    YES: ⇒ 3   No: **STOP**   COMMENT: EPITHETS MUST BE PROPERLY PUBLISHED.

3   After 1958, was the publication dated at least to the year? (Art. 21.6)
    YES: ⇒ 4   No: **STOP**   COMMENT: REJECT AS NOT ESTABLISHED
                                PUBLICATIONS SINCE 1958 MUST BE DATED - FIND A
                                LATER PUBLICATION THAT IS DATED.

4   Is there a statement of the component genera of the graft-chimaera? (Art. 20.3)
    YES: ⇒ 5   No: **STOP**   COMMENT: REJECT AS NOT ESTABLISHED.

THE NAME OF THE GRAFT-CHIMAERA IS ESTABLISHED

5   Is the name formed from accepted generic names?
    YES: ⇒ 6   No: **STOP**   COMMENT: REJECT AS NOT ACCEPTABLE AND USE A
                                NAME FORMED FROM ACCEPTED GENERIC NAMES.

6   Is there an earlier established name for the graft-chimaera?
    No: ⇒ 7   YES: **STOP**   COMMENT: REJECT AS NOT ACCEPTABLE AND USE THE
                                EARLIER ESTABLISHED NAME.

7   <u>YOU HAVE THE ACCEPTED NAME FOR THE GRAFT-CHIMAERA</u>

APPENDIX VIII

QUICK GUIDE FOR NEW CULTIVAR NAMES

Many of the Rules of this *Code* deal with sorting out problems which have arisen in past nomenclature. The following notes are intended as a quick guide to forming new cultivar epithets and should be read by everyone wishing to name a new cultivar.

DO I HAVE A NEW CULTIVAR?

You have a new cultivar and you wish to name it. First check that you do actually have a cultivar. A single plant is not a cultivar: a cultivar is a group of individual plants which collectively is distinct from any other, which is uniform in its overall appearance and which remains stable in its attributes. Do not attempt to name a cultivar until you have a number of individuals which are uniform and stable. Now convince yourself that your cultivar is really worth naming; there is no point in going through the process of naming your cultivar if it is not an improvement on others.

There are different sorts of cultivar ranging from clones which should be genetically identical to tightly–controlled seed–raised cultivars such as F_1 hybrids. Article 2 of this *Code* defines the different kinds of cultivar.

The only way you can check if it new is by comparison with existing cultivars. Your new cultivar must be distinguishable from those already in existence.

Once you are satisfied that you do indeed have a new cultivar, decide if you want to give it a cultivar epithet. The epithet is the last part of the entire name which renders the name unique. Cultivar epithets are always written within single quotation marks so that they stand out from the rest of the name.

Remember that cultivar epithets, by their very definition, are available for all to use and that the epithets themselves offer no protection if you wish to obtain intellectual property rights on your new cultivar.

HOW DO I FORM A NEW CULTIVAR NAME?

The full name of a cultivar will always begin with the botanical name of the genus to which the cultivar belongs. Optionally, the species or hybrid may be included as a second element in the entire name but this is not necessary; inclusion merely provides more information about your cultivar.

Nowadays, new cultivar epithets must be in a modern language and they must be unique within the so-called denomination class which is usually the genus. Some groups have special denomination classes and these may be found in Appendix IV of this *Code*.

Coining a new and original cultivar epithet is not easy, especially in groups which historically have had hundreds or even thousands of cultivars. Luckily many of these groups have International Registration Authorities (IRAs) who publish check-lists and registers of epithets which have been used in the past. Check in Appendix II in this *Code* to see if the genus of your cultivar is covered by an IRA and then consult the IRA's publications. Each IRA has a registrar who will be glad to advise you if your proposed epithet has been used before and whether or not your epithet is acceptable.

There have been many other lists of cultivar epithets produced in the past and a fairly comprehensive list of those is given in Appendix XI of this *Code*. Most good horticultural and scientific libraries will have copies of these publications for you to check for prior publication.

Composing an epithet requires a bit of thought. An ideal epithet is both easy to spell and pronounce in the various countries that the cultivar might be distributed. The rules for composing an epithet allow you to use or make up any word or words you want but the epithet will not be allowed as a cultivar epithet if it is confusing or likely to confuse. This *Code* governs the reasons why a proposed epithet might not be allowed; disallowed epithets are to be "rejected".

The following is a check list of things to do:

1　　　　Make sure your proposed epithet is unique.

2　　　　Make sure that your epithet cannot be confused either in spelling or pronunciation with another existing one.

3　　　　Make sure that your epithet could not be interpreted as being likely to exaggerate the merits of the cultivar ('Best Ever', 'The Greatest' and 'Tastiest of All' are not acceptable for obvious reasons!)

4　　　　Make sure that your epithet has no more than 10 syllables and no more than 30 characters, excluding spaces and the single quotation marks.

5　　　　Make sure that your epithet is not only made up of simple descriptive words like 'Red', 'Giant White' or 'Small'.

6　　　　Do not use any of the following banned words or their equivalents in any language in your epithet: "cross", "hybrid", "grex", "group", "form", "maintenance", "mutant", "seedling", "selection", "sport", "strain", "variety" (or the plural form of these words in any language) or the words "improved" or "transformed".

7   Do not use any punctuation marks except for the apostrophe, the comma, a single exclamation mark, the hyphen and the full-stop.

8   If your epithet is a single word, make sure that the word is not the same as that of a genus, whether in botanical Latin or in a modern language. (Erica, Daphne, Iris and Veronica happen to be Latin names of genera and are not permitted as cultivar epithets even though they are personal names as well. Rose and violet are common names of genera and they too are not permitted; these would not be acceptable under 5 above anyway.)
Such a word may be used in an epithet of two or more words provided that it does not form the final word. ('Erica Smith', 'Iris Jones' and 'Rose Queen' are acceptable.)

9   Make sure that your epithet does not contain the botanical or common name of its genus or the common name of any species in that genus. (*Rosa* 'Christmas Rose', Potato 'Jim's Spud' and *Primula* 'White Cowslip' are not acceptable.)

WHAT DO I DO WITH MY NEW NAME?

Once you have satisfied yourself that your epithet is in an acceptable form, register it with the appropriate IRA. This will cost you little more than the time spent filling in a form and sending it off but will help ensure that the epithet is internationally recognized forever.

The epithet will have to be published in order to be absolutely fixed. You may either publish it yourself, say in your nursery catalogue if you are a nurseryman, or the IRA concerned will publish it for you in due course if you register the epithet with them. IRAs, however are placed under no obligation to publish your epithet within a short period of time and you should realise that your chosen epithet might be used by someone else for a completely different plant unless you take steps to ensure early publication. If someone else, even if in a different part of the world, publishes your chosen epithet for a different cultivar in the same genus, you will have to think of another.

Publication of your new epithet must be in printed or similarly duplicated matter which is distributed to the general public or at least to botanical, agricultural, forestry or horticultural institutions with libraries. News-papers, non-technical magazines and similar publications which are not designed to last do not count as publications in this case.

Publications must be dated. A new epithet appearing in a nursery catalogue will not be treated as having been published if that catalogue is undated at least to the year.

Do not publish more than one epithet for the same cultivar in the same publication: if you do this none will be considered as having been published in that publication.

It may be that you are registering or publishing a new cultivar epithet on behalf of someone else or that you are promoting a new epithet for a cultivar raised by someone else. Check that the originator of the cultivar agrees with the epithet (and its spelling) that you are promoting; if he does not, the epithet may have to rejected in favour of the originator's choice.

When you publish a new cultivar epithet, you must include a description of the cultivar. The longer and more complete the description the better, but at least state its obvious characteristics and if you can, state how it differs from an existing cultivar. It is helpful, but not compulsory, to supply an informative illustration of the new cultivar in the publication if expense permits.

Make a statement such as "new cultivar name" after the new epithet so that others may recognize the fact that you have deliberately named it.

HOW CAN I PROTECT MY NEW NAME?

Send a copy of your publication to the IRA and to the main horticultural libraries in your part of the world. If you are feeling generous, send copies to similar libraries in other parts of the world too.

If you can, distribute herbarium specimens (Standards) of the new cultivar to as many herbaria as is practical but certainly to your nearest herbarium which specializes in maintaining Standards. (A list is provided in Appendix VI.) This will help ensure that your cultivar will not become confused with others in the future and may help resolve disputes if more than one person thinks they have raised the same cultivar!

Finally, ensure that the name is used by everyone and do not encourage others to coin trade-designations or other selling names for your plant. The most effective way to protect a name is to label your plants clearly and unambiguously. Always maintain "your" cultivar epithet within single quotation marks to ensure that the status of your plant is understood.

APPENDIX IX

LATIN NAMES OF PLANTS

The formation and use of botanical names of plants in Latin form are governed by the *International Code of Botanical Nomenclature*, also called the Botanical Code or *ICBN*. The current edition (1994) was formulated as a result of decisions taken at the Fifteenth International Botanical Congress held in Japan in August–September 1993 and is commonly known as the Tokyo Code.

The *ICBN* deals with the naming of taxonomic groups (abbreviated to taxa: singular taxon) and these are arranged in a hierarchical order in consecutively subordinate ranks.

The principle ranks of taxa in descending sequence are: kingdom, division (or phylum), class, order, family, genus and species of which the species is considered the basic rank. Only the final three ranks are in general use for the nomenclature of cultivated plants.

FAMILY NAMES
These end with the letters –*aceae* and are based upon the name of a genus.

> Examples: *Apiaceae* based on the genus *Apium*, *Brassicaceae* based on *Brassica*, *Caryophyllaceae* based on *Caryophyllus*, *Ginkgoaceae* based on *Ginkgo*, *Rosaceae* based on *Rosa*, *Winteraceae* based on *Wintera*.

An exception to this rule is made for long–standing family names which are descriptive and are not based on generic names; their use is authorized as alternative names.

> Examples: *Apiaceae* is based on the genus *Apium*, but the long–standing name *Umbelliferae* is permitted as an alternative; likewise *Poaceae*, based on *Poa* is better known by some as *Gramineae*.

GENUS NAMES
These are a substantive (noun) in the singular, or a word treated as such, and are written with a capital initial letter.

> Examples: *Faba, Lilium, Pinus, Triticum, Zea,* ×*Crataemespilus,* +*Crataego-mespilus*.

The name of a subdivision of a genus, such as subgenus, sectio, or series, is a combination of a generic name and a subdivisional epithet connected by a term denoting its rank. The epithet is written with a capital initial letter and when written in connection with a specific epithet, is placed in parentheses.

> Examples: *Prunus* (subg. *Cerasus*) *avium*; *Primula* (sect. *Candelabra*) *japonica*; *Iris* (ser. *Laevigatae*) *laevigata*.

SPECIES NAMES

The scientific name of a species is a binominal combination (binomen) in Latin form consisting of the name of the genus followed by a single specific epithet.

> Example: *Lilium candidum*, where *Lilium* is the generic name and *candidum* the specific epithet.

Specific epithets (sometimes also called trivial names), when adjectival in form, agree in gender with the name of their genus. They are written with a lower-case initial letter.

The name of a subdivision of a species, such as subspecies, varietas, and forma, is a combination of a specific name and a subspecific epithet connected by a term denoting its rank.

> Examples: *Ranunculus acris* subsp. *friesianus*; *Rosa sericea* var. *omeiensis*.

VALID PUBLICATION

To be validly published (established, using the terminology of this *Code*), names of taxa have to formed in accordance with the provisions of the *ICBN*, be properly published with a Latin description or diagnosis and must have a name-bearing type designated to which the name is permanently attached, whether it is a correct name or a synonym.

HYBRIDS

The Rules for naming hybrids is covered in Appendix I of the *ICBN* (the "Hybrid Appendix"). Hybridity is indicated by the use of the multiplication sign ×, or by adding the prefix "notho-" (from the Greek *nothos*, meaning hybrid) to the name of the rank of the taxon e.g., nothogenus, nothospecies.

Hybrid formulae

A hybrid between named taxa (a nothotaxon) may be indicated by placing the multiplication sign between the names of the taxa; such an expression is called a hybrid formula.

> Examples: the hybrid formula for crosses between the grasses *Agrostis* and *Polypogon* is written *Agrostis* × *Polypogon*; the hybrid formula for crosses between *Camellia japonica* and *C. saluenensis* is *Camellia japonica* × *Camellia saluenensis*.

The order of the names in a hybrid formula may be either alphabetical (as in this *Code*), or, when the female parent is known, with the name of the female parent first. The male (♂) and female (♀) signs may be added if desired. The method used throughout a particular publication should be clearly stated.

When a hybrid formula is used with a cultivar epithet, the hybrid formula should be given in parentheses before the epithet.

> Example: the camellia cultivar 'Donation', may be designated (*Camellia japonica* × *C. saluenensis*) 'Donation'.

Names of hybrids
Hybrids between representatives of two or more taxa may receive a name. The hybrid nature of a taxon is demonstrated by the addition of the multiplication sign × before the designation of nothogenus or the epithet of a nothospecies.

> Examples: using the above examples, the name of the nothogenus *Agrostis* × *Polypogon* is ×*Agropogon*; the name of the nothospecies *Camellia japonica* × *Camellia saluenensis* is *C.* ×*williamsii*.

The multiplication sign in the name of a hybrid should be placed against the initial letter of the name or epithet. However, if the mathematical symbol is not available and the letter "x" is used instead, a single letter space may be left between it and the epithet if this helps to avoid ambiguity. The letter "x" should be in lower case.

All members of a nothogenus, whatever the species, have the same nothogeneric name.

> Example: all progeny derived from the crossing of any *Cupressus* species with any *Chamaecyparis* species may be designated by the nothogenus ×*Cupressocyparis*.

Formation of nothogeneric names
The nothogeneric name for a cross involving two genera is formed by a combination of parts of the names of the two parent genera, using the first or whole of one, the last part or the whole of the other (but not the whole of both) and, optionally, a connecting vowel. Such names are termed condensed formulae.

> Examples: ×*Amarcrinum* for *Amaryllis* × *Crinum*, ×*Gymnanacamptis* for *Anacamptis* × *Gymnadenia* and ×*Mahoberberis* for *Berberis* × *Mahonia*.

The nothogeneric name for a cross involving four or more genera is formed from the name of a person to which is added the termination *–ara*. Such a name is regarded as a condensed formula but must not exceed eight syllables.

> Example: ×*Potinara* is the condensed formula for *Brassavola* × *Cattleya* × *Laelia* × *Sophronitis*.

The nothogeneric name for a cross involving three genera is either a condensed formula in which the names of the three parental genera are combined into a word not exceeding eight syllables, using the whole or first part of one, followed by the whole or any part of another, followed by the whole or last part of the third (but not the whole of all three) and, optionally, one or two connecting vowels, or it is formed from the name of a person to which is added the termination *–ara*.

When a nothogeneric name is formed from the name of a person, it is the tradition that that person should be a collector, grower or student of the group concerned.

In order to be validly published (established), the name of a nothogenus must be published with a statement of the names of the parent genera, but no Latin description or diagnosis is necessary. Since the names of nothogenera are condensed formulae, or are treated as such, they do not have name–bearing types.

Formation of nothospecific names
The name of a nothospecies is formed and validly published (established) in the same way as that for a species; the hybrid nature is indicated by the addition of the multiplication sign before the epithet. It is not compulsory to use the multiplication sign, which is merely added before the epithet to indicate extra information about the status of the species.

All progeny from a particular combination of species have the same nothospecific epithet.

> Examples: all progeny derived from crossing *Cupressus macrocarpa* and *Chamaecyparis nootkatensis* may bear the nothospecific name ×*Cupressocyparis leylandii*; progeny derived from *Cupressus glabra* and *Chamaecyparis nootkatensis* may bear the name ×*Cupressocyparis notabilis*; *Lilium* ×*sulphurgale* is the name for hybrids between *Lilium regale* and *L. sulphureum*.

REMINDER OF RECOMMENDATION 16A OF THIS *CODE*:
Names of taxonomic groups of cultivated plants below the rank of genus should be named in accordance with the provisions of this *Code*, and not under the provisions of the *ICBN*.

APPENDIX X

FLOW CHART OF NOMENCLATURAL BODIES AND PROCESSES

This chart demonstrates the relationships between the various international bodies that, in effect, govern the international nomenclature of plants, resulting in the provision of a correct nomenclature.

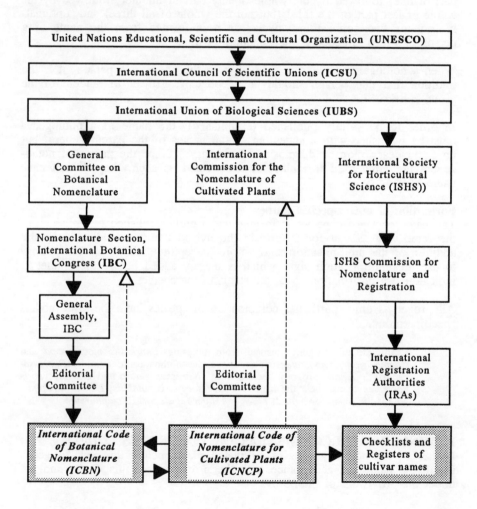

APPENDIX XI

CHECKLISTS OF ORNAMENTAL CULTIVARS

This Appendix is taken from *Arnoldia*[1] (Winter 1994–1995) Vol. 55, No. 4. (1995) © The President & Fellows of Harvard College.

The Editorial Committee is most grateful to the Arnold Arboretum of Harvard College, Karen Madsen the editor of *Arnoldia*, and Professor Tucker and his colleagues for permission to include this list of checklists which will be an invaluable reference for those wishing to check precedence, establishment and acceptability of cultivar names.

An extensive bibliography follows the list of checklists.

It is hoped that additions, corrections and updates to this list will be included in future editions of this *Code*.

INTRODUCTION TO THE LIST OF
CHECKLISTS OF ORNAMENTAL CULTIVARS

This list originated in an attempt to locate pre–1900 cultivars of ornamentals for historical restoration work and to verify them by cumulative checklists (Kunst and Tucker, 1989). This goal was expanded in an attempt to locate all cultivar checklists for ornamental plants. Since these lists are often scattered and sometimes difficult to locate, the compilers make no claim that the list is complete.

The following lists some historical checklists but concentrates on the most recent, updated, cumulative ones. The cut–off date for inclusion, with few exceptions, is January 1994.

Since the ideal of providing coloured illustrations in checklists is rarely achieved, especially in the older literature, this list also includes a number of taxonomic and other botanical and horticultural revisions when cultivars are described and illustrated. Good general references to extant cultivars are Harkness and D'Angelo (1986) and Hatch (1986). The reports of Wisley Trials in the *Journal* and *Proceedings of the Royal Horticultural Society* are recommended for cultivar descriptions, while Wright (1984) also discusses many cultivars.

Besides numerous cultivar names in the "Cultivar & Germplasm Releases" section, *HortScience* has become the vehicle for publication of cultivar names for unassigned woody genera (Huttleston, 1986, 1988, 1989, 1990, 1991, 1992, 1993; Vrugtman, 1994), genera at the Arnold Arboretum (Spongberg, 1988, 1989, 1990,

---

[1] *Arnoldia* (ISBN 004–2633; USPS 866.100) is published quarterly by the Arnold Arboretum, 125 Arborway, Jamaica Plain, Massachutsetts 02130 3519, United States of America

1991, 1992, 1993, 1994a), *Kalmia* (Jaynes, 1989), and *Syringa* (Vrugtman, 1988, 1989a, 1989b, 1990a, 1991, 1994b). Listings of other cultivar names are supposed to be maintained and published by International Registration Authorities (American Association of Nurserymen, c. 1987; Leslie, 1986; Schneider, 1986a; Vrugtman, 1972, 1973, 1977, 1981, 1984, 1985, 1986, 1989c, 1990b, 1990d). The work of these designated authorities is further supplemented by cumulative checklists and origination lists maintained and published by specialist societies.

Notice of any corrections, updates and additions would be greatly appreciated and should be sent to one of the authors listed below.

Arthur O. Tucker, Department of Agriculture and Natural Resources, Delaware State University, Dover, Delaware 19901-2277, United States of America

Scott G. Kunst, 536 Third Street, Ann Arbor, Michigan 48103, United States of America

Freek Vrugtman, Royal Botanical Gardens, Box 399, Hamilton, Ontario L8H 3H8, Canada

Laurence C. Hatch, P. O. Box 12011, Raleigh, North Carolina 27605, United States of America

LIST OF CHECKLISTS OF ORNAMENTAL CULTIVARS

Abies See conifers.

Acacia See Australian & South African plants.

Acer Cultivars of maples are listed by Bean (1970–1988), Bom (1982), Grootendorst (1969a), Harris (1983), Krüssmannn (1984–1986), Mulligan (1958), Murray (1970), Schwerin (1919), and Weaver (1976b). The cultivars of the vine maple (*A. circinatum* Pursh) are discussed by Vertrees (1979). Cultivars of red maple (*A. rubrum* L.), sugar maple (*A. saccharum* Marsh.), Norway maple (*A. platanoides* L.), and silver maple (*A. saccharinum* L.) are listed by Santamour & McArdle (1982a, 1982b, 1982c, 1982d). The cultivars of Japanese maple (*A. palmatum* Thunb.) are thoroughly documented by Harris (1982) & Vertrees (1978, 1987) with color photographs. Additional recent registrations are recorded in Huttleston (1986, 1989).

Achimenes See *Gesneriaceae*.

Aconitum Some of the cultivars of the monkshoods are listed by Lord (1988) and Müssel (1986).

Adiantum See ferns.

Adonis Cultivars of *Adonis amurensis* Regel & Radde are discussed by Nakamura (1964).

Aeschynanthus See *Gesneriaceae*.

Aesculus The cultivars of the horsechestnuts are discussed by Bean (1970–1988), Grootendorst (1967a), Krüssmannn (1984–1986), and Wright (1985).

Agonis See Australian & South African plants.

Agapanthus The cultivars of the Nile lilies are discussed in the Wisley Trials of 1977 (Royal Horticultural Society, 1978).

Aglaonema The cultivars of the Chinese evergreens are listed by Jervis (1980).

Allium The few ornamental cultivars of *Allium,* the onions, are listed by Davies (1992).

Alnus Cultivars of the alders are listed by Ashburner (1986) but without introduction dates or background. Grootendorst (1972a) and Schneider (1965a) also discuss the cultivars of alders. An additional registration of *Alnus* is recorded by Huttleston (1988). Cultivars of *Alnus* are also discussed by Bean (1970–1988) and Krüssmann (1984–1986).

Aloe The South African Aloe Breeders Association has circulated an unpublished list of *Aloe* cultivars (for example, 1987), and many cultivars are published in *Aloe* and other South African journals.

Alsobia See *Gesneriaceae.*

Alyssum The cultivars of the alyssums are discussed by Dudley (1966).

Amaryllidaceae Traub & Hannibal (1960) list the cultivars of *Brunsvigia* with later additions published in *Plant Life.* Traub (1961) lists the cultivars of x *Crinodonna* with later additions in *Plant Life.* Kelsey & Dayton (1942) and Anonymous (1958f) are the first to list the cultivars of the garden amaryllis, *Hippeastrum,* but the most comprehensive list is by Traub et al. (1964) with subsequent registrations in *Plant Life.* The nerines are listed by Kelsey & Dayton (1942), the Royal General Bulbgrowers' Association (1991), and Smithers (1993), but the most comprehensive lists are Menninger (1960), Roberts (1984), and Smee (1984) with later additions in editions of *Plant Life.*

Amelanchier A checklist of ornamental and fruiting shadbush cultivars is Hilton (1982, 1984). Krüssmann (1984–1986) also lists cultivars.

Anigozanthos See Australian & South African plants.

Anemone Many cultivars of *Anemone* are listed by Trehane (1989). The cultivars of *A. nemorosa* L. are listed by the Royal General Bulbgrowers' Association (1991) and Toubøl (1981). The history and performance of cultivars of *Anemone japonica* (Thunb.) Sieb. & Zucc. are discussed by Clausen (1972a) and Hensen (1968, 1979).

Antirrhinum The cultivars and performance of the snapdragons are listed by the Royal Horticultural Society (1913b).

Arctostaphylos The cultivars of the bearberries and manzanitas are recorded by Keeley & Keeley (1994).

Ardisia The Japanese cultivars of *Ardisia japonica* (Hornst.) Blume, the marlberry, are discussed by Yinger & Hahn (1985).

Argyranthemum The cultivars of the marguerite are compared and contrasted with studio photographs by Cheek (1993).

Aster The most comprehensive lists of the cultivars of the asters are by Meier (1973a, 1973b, 1973c, 1973d) and Jensma (1989); the latter is being expanded and revised. Kelsey & Dayton (1942), Royal Horticultural Society (1902, 1908a, 1926a), and Trehane (1989) discuss the cultivars of the hardy asters, but these are published without introduction dates or background. Ranson (1946) lists mostly species with few cultivars. The history and performance of cultivars of the asters are discussed by Allen (1983), Clausen (1973a), and Jelitto & Schacht (1990). Barret (1959) discusses the performance of cultivars of *A. ericoides* L.

Astartea See Australian & South African plants.

Astilbe The most comprehensive published list of *Astilbe* cultivars is Ievinya & Lusinya (1975) with c. 170 cultivars, detailed descriptions of c. 50, and an extensive bibliography. Hensen (1969) discusses the history and performance of species and cultivars of *Astilbe.* Jelitto & Schacht (1990), the Royal Horticultural Society (1970b), Schneider (1968), and Trehane (1989) also list cultivars of *Astilbe.*

Aubrieta The cultivars of *A. columnae* Guss., *A. deltoidea* (L.) DC., and *A. intermedia* Heldr. & Orph. are thoroughly discussed by Clausen (1973c) and

Hensen (1976). Jelitto & Schacht (1990) and Trehane (1989) also list cultivars of *Aubrieta.*

Australian & South African plants The Australian Cultivar Registration Authority has circulated a continually updated list (for example, 1988) of registered cultivars of *Acacia, Agonis, Anigozanthos, Astartea, Baeckea, Banksia, Baura, Blechnum, Boronia, Brachychiton, Brachycome, Callistemon, Callitris, Ceratopetalum, Chamelaucium, Correa, Crowea, Epacris, Eremophila, Eriostemon, Eucalyptus, Eucryphia, Grevillea, Hakea, Hardenbergia, Helichrysum, Hypocalymma, Kennedia, Kunzea, Lechenaultia, Leptospermum, Lophostemon, Melaleuca, Myoporum, Pandorea, Pimelea, Plectranthus, Prostanthera, Pultenaea, Scaevola, Spyridium, Telopea,* and *Tetratheca.*

Baeckea See Australian & South African plants.

Banksia See Australian & South African plants and *Proteaceae.*

Baura See Australian & South African plants.

Begonia The most comprehensive checklist of begonias is Ingles (1990). This should be supplemented with Kelsey & Dayton (1942), Thompson (1976–1978, 1984), and Thompson & Thompson (1980, 1982). Cultivars of *B. semperflorens–*cultorum hybrids are published by Maatsch (1962), Maatsch & Nolting (1969, 1971a), and Nolting & Zimmer (1975a, 1980a, 1985, 1987). Cultivars of tuberous begonias are listed by Haegeman (1978, 1979) and Langdon (1969). Cultivars of other begonias are published by the American Begonia Society (1957, 1958, 1962, 1967, 1985). Registrations in *The Begonian* are summarized by Vrugtman (1972). These should be used in conjunction with Japan Begonia Society (1980), Misono (c. 1974–1978), and Thompson & Thompson (1981).

Berberis Schneider (1923) covers the publication history and descriptions of cultivars of the barberries. The cultivars of the barberries are also listed by Ahrendt (1942, 1949, 1961), Krüssmann (1984–1986), Laar (1972), and Wyman (1962b).

Bergenia The cultivars of the bergenias are discussed by Beckett (1983), Laar (1973), and Yeo (1971a, b).

Betula Cultivars of the birches are listed by Ashburner (1980), Fontaine (1970a), Grootendorst (1973a), Jong (1986), Santamour & McArdle (1989), and Wyman (1962c). Cultivars of *Betula* are also discussed by Bean (1970–1988) and Krüssmann (1984–1986).

Blechnum See ferns.

Boronia See Australian & South African plants.

Bougainvillea The most comprehensive discussions of cultivars of bougainvillea are Choudhury & Singh (1981) and MacDaniels (1981). Previous checklists are Anonymous (1959c) and Gillis (1976). Subsequent registrations are recorded by Singh (1986).

Brachychiton See Australian & South African plants.

Brachycome See Australian & South African plants.

Bromeliaceae A preliminary checklist of bromeliad cultivars has been prepared by Beadle (1991) and the Bromeliad Society (1989).

Brunsvigia See *Amaryllidaceae.*

Bucinellina See *Gesneriaceae.*

Buddleja The cultivars of the butterfly bushes are listed by Leeuwenberg (1979) but without dates of introduction or hybridizers. Cultivars of the butterfly bushes are also discussed by Bean (1970–1988), Grootendorst (1972b), and Krüssmann (1984–1986).

Bulbs The cultivars of many hardy and tender bulbs are published by the Royal General Bulbgrowers' Association (most recent is 1991). Many cultivars are also listed in Trehane (1989).

Buxus A guide for registration and documentation of cultivar names of *Buxus* is provided by Dudley & Eisenbeiss (1971). The cultivars of boxwood are listed by Bean (1970–1988), Batdorf (1987, 1988), Krüssmann (1984–1986), and Wagenknecht (1965, 1967, 1971, 1972).

Cactaceae Cultivars of many succulents, including cacti, are listed by Jacobsen (1977). The cultivars of *Epiphyllum*, the orchid cacti, are thoroughly discussed by Rainbow Gardens (1979), along with cultivars of *Epiphyllum* x *Aporocactus*. This should be supplemented with Hashizume (c. 1982–1985), who provides good color photographs and English captions in his guide to taxa of *Epiphyllum;* additional color photographs are presented by Leue (c. 1987). The cultivars of *Schlumbergera* are discussed by Horobin (1985).

Caladium The cultivars of the caladiums are evaluated for landscaping by Wilfret (1984).

Callistemon See Australian & South African plants.

Callistephus The cultivars of the China aster, *C. chinensis* (L.) Nees, are assembled into checklists by Maatsch (1958, 1964), Maatsch & Nolting (1971c), Nolting & Zimmer (1975c, 1981, 1987), and Olmsted et al. (1923).

Callitris See Australian & South African plants.

Calluna A guide to naming heather cultivars is McClintock (1986). Bean (1970–1988), Chapple (1951), Johnson (1956), Krüssmann (1984–1986), Laar (1968, 1970a, 1974, 1977a), Letts (1966), Proudley & Proudley (1974), and Underhill (1990) list many cultivars of heathers. Munson (1981, 1984) provides a comprehensive key to the species and cultivars of *Calluna* with full botanical descriptions in the 1981 thesis.

Camellia Bean (1970–1988), Durrant (1982), Erdman (1949), Gerbing (1945), Hertrich (1954–1959), Hume (1955), Krüssmann (1984–1986), Macoboy (1981), and Sharp (1957) list many cultivars of camellias, but Woodroff & Donnan (1990) is probably the best compact checklist, while Savige (1993) is probably the most complete list with 41,000 cultivars. The Japanese cultivars of camellias are listed by Tuyama (1968), while the Chinese cultivars are listed by the Kunming Institute of Botany, Academica Sinica (1986). The International Camellia Society expects to publish the International Camellia Register in 1992.

Campanula The history and performance of cultivars of the bellflowers are discussed by Clausen (1976) and Lewis & Lynch (1989). Trehane (1989) also lists the cultivars of the bellflowers.

Canna Kelsey & Dayton (1942) has published a list of canna cultivars without dates and hybridizers. Additional information is given by the Royal General Bulbgrowers' Association (1991), Royal Horticultural Society (1908b, 1909), and Trehane (1989). Mukherjee & Khoshoo (1970) provide botanical characteristics of many cultivars.

Capsicum The peppers are sometimes grown as ornamentals (e.g., 'Fips'), and Andrews (1984) records extensive information on these cultivars in her monumental book.

Carnivorous plants Named cultivars of *Drosera, Nepenthes,* and *Sarracenia* are listed by Schlauer (1986, 1987; note that the first list neglects to capitalize the cultivars) and Kusakabe (1987). Additional cultivars of *Sarracenia* are later listed by Mellichamp & Gardner (1987). The hybrids of *Nepenthes* are reported by Fleming (1979). Fleming's list is reprinted in Pietropaolo & Pietropaolo (1986). An additional cultivar of *Nepenthes* is listed by Robinson (1989).

Carpinus Cultivars of *Ostrya*, the hop hornbeams, and *Carpinus*, the hornbeams, are discussed by Rushforth (1985), Schneider (1965a), and Wright (1986). Cultivars of *Carpinus* and *Ostrya* are also discussed by Bean (1970–1988) and Krüssmann (1984–1986).

Castanea The cultivars of the potentially blight–resistant chestnuts are discussed by Jaynes & Graves (1963) and Nienstaedt & Graves (1955).

Ceanothus Van Rensselaer & McMinn (1942) provide the most comprehensive listing of the wild–lilacs and buckrushes. Additional cultivars are listed by Bean (1970–1988), Hogan (1988), Huttleston (1986), Keeley & Keeley (1994), Krüssmann (1984–1986), Schmidt (1962), and Smith (1979).

Ceratopetalum See Australian & South African plants.

Cercis The cultivars of the redbuds are discussed by Raulston (1990).

Chaenomeles The cultivars of the Japanese quinces are listed by Bean (1970–1988), Grootendorst (1968a), Krüssmann (1984–1986), and Weber (1963).

Chamelaucium See Australian & South African plants.

Chrysanthemum See *Argyranthemum* for the marguerite and *Dendranthema* for the florist's chrysanthemum.

Cistus The cultivars of the rock roses are discussed by Bean (1970–1988), Warburg (1931), and Warburg & Warburg (1930).

Citrus The cultivars of *Citrus*, some ornamental, are listed by Hodgson (1967).

Clematis A general clematis checklist is Lloyd (1965, 1989); Fretwell (1989) provides good color photographs. The cultivars of *C. viticella* L. are listed by Rogerson (1985). The large–flowered clematis hybrids are published by Evison (1985) and Spingarn (1935), while the hybrids of section *Atragene* are published by Pringle (1973). Kelsey & Dayton (1942) also provide a list of *Clematis* but without introduction dates and background. Some cultivars are also published by Laar (1985), Markham (1935), and Trehane (1989).

Codiaeum The list for the garden croton, *Codiaeum variegatum* (L.) Blume, of Kelsey & Dayton (1942) provides no introduction dates or background. Additional cultivars are listed by Anonymous (1959a) and Brown (1960). The latter provides good color illustrations but also includes a number of botanical errors.

Coleus–Plectranthus While no proper checklist exists for coleus cultivars, Pedley & Pedley (1974) and Stout (1916) provide many materials for the production of such a list. Recent registrations of *Plectranthus* are listed by the Australian Cultivar Registration Authority (1988).

Columnea See *Gesneriaceae*.

Conifers Lewis (1986a, b) provides a guide for the naming of conifer cultivars. The Royal Horticultural Society has started an ambitious registry of conifers; so far *Abies* to *Pherosphaera* have been published (Lewis, 1985; Lewis & Leslie, 1987, 1989). Meanwhile, Welch & Haddow (1993) have published a world checklist of conifers. Den Ouden & Boom (1978) and Krüssmann (1985) present the cultivated conifers with introduction dates and descriptions, while Cope (1986) provides scant information on dates or origins. Welch (1991) is an updating of both den Ouden & Boom (1978) and Welch (1979). Many cultivars of conifers are also listed by Bean (1970–1988). Munson (1973) has prepared a vegetative key to dwarf and slow–growing conifers, while Obrizok (1991) provides growth forms of dwarf conifers. The cultivars of the firs (*Abies* spp.) and spruces (*Picea* spp.) are listed by Gelderen (1975). Silver firs, *Abies alba* Mill., are discussed by Horsman (1984). Japanese cedar, *Cryptomeria japonica* (L. f.) G. Don, cultivars are treated by Kortmann (1987) and Tripp (1993). Juniper (*Juniperus* spp.) cultivars are listed by Gelderen (1984) and Grootendorst (1968c). Pine (*Pinus* spp.) cultivars are listed by Gelderen (1982), while the cultivars of

Japanese five needle pine, *P. parviflora* Sieb. & Zucc., are listed by Valavanis (1976). The cultivars of yews, *Taxus*, are listed by Chadwick & Keen (1976). The cultivars of American arborvitae, *Thuja occidentalis* L., are listed by Grootendorst (1971b), while Wyman (1961c) discusses cultivars of four arborvitae species. Hemlock (*Tsuga* spp.) cultivars are documented in Swartley (1984). These should be supplemented with the color photographs of Harrison (1975) and Gelderen & van Hoey Smith (1986) and the black and white photographs of Welch (1979).

Cordyline   A list for the cultivars of the ti, *Cordyline terminalis* (L.) Kunth, is Kelsey & Dayton (1942) but without introduction dates and background. Additional cultivars are listed by Anonymous (1959d).

Cornus   Cultivars of the dogwoods are listed by Bean (1970–1988), Howard (1961), Krüssmann (1984–1986), and Santamour & McArdle (1985a). Jaynes, Brand, & Arnow list the cultivars of the kousa or Japanese dogwood, *C. kousa* Hance. Additional registrations are recorded by Spongberg (1988, 1990).

Correa   See Australian & South African plants.

Corylus   Bibliographic references to the cultivars of the filberts are recorded by Debor (1978).

Cosmos   The Indian cultivars of cosmos are listed by Anonymous (1959b).

Cotoneaster   The cultivars of the rock sprays are treated by Bean (1970–1988), Grootendorst (1966a), Hachmann et al. (1987), and Krüssmann (1984–1986).

Crataegus   Wyman (1962d) lists the cultivars of the hawthorns but with few introduction dates and little background. Bean (1970–1988), Grootendorst (1967c), and Krüssmann (1984–1986) also discuss the cultivars of the hawthorns.

× *Crinodonna*   See *Amaryllidaceae*.

Crinum   The cultivars of the spider lilies are listed by Hannibal (1970–71).

Crocosmia   The cultivars of the montbretias are listed by Kostelijk (1984).

Crocus   The cultivars of crocus are documented by the Royal General Bulbgrowers' Association (1991), but further information is provided by Ruksans (1981) and Trehane (1989).

Crowea   See Australian & South African plants.

Cryptomeria   See conifers.

Cyclamen   Some cultivars of *Cyclamen* are recorded by Grey–Wilson (1988), the Royal General Bulbgrowers' Association (1991), Trehane (1989), and Wellensiek (1961), while a comprehensive list of cultivars is Wellensiek et al. (1961).

Cytisus   The cultivars of the brooms (*Cytisus* and *Genista)* are treated by Bean (1970–1988), Krüssmann (1984–1986), and Laar (1971).

Daboecia   *Daboecia* cultivars are included in many listings of heaths and heathers (e.g., Johnson, 1956; Munson, 1981; Underhill, 1971), and separately by Bean (1970–1988), Krüssmann (1984–1986), and Laar (1977b).

Dahlia   The most comprehehensive checklist of dahlias is that of the Royal Horticultural Society (1969a) and later supplements (1988d, 1989c, 1992a). Unfortunately, with one exception (a cultivar from 1850), this list omits all cultivars 1789–1859 and many of the cultivars 1860–1900. Many early twentieth–century cultivars are amply covered in Norton (1924), Olmsted et al. (1923), and Sandhack (1927). Recent cultivars are listed by the American Dahlia Society (e.g., 1989) in a paperbound booklet as a supplement to the *Bulletin of the American Dahlia Society*.

Dalbergaria   See *Gesneriaceae*.

Daphne   The cultivars of *Daphne* are listed by Brickell & Mathew (1976). Bean (1970–1988), Hodgkins (1961), and Krüssmann (1984–1986) also discuss the cultivars of *Daphne*.

Delphinium The Royal Horticultural Society (1970a) is a checklist of delphinium
   names that updates a list previously published in 1949. International registration
   of delphinium cultivars is handled by the Delphinium Society (Cooper, 1984,
   1986, 1987, 1989, 1990b, 1991, 1992, 1993, 1994). Edwards (1987) and Trehane
   (1989) are recent lists of cultivars. Bishop (1949), Cooper (1990a), Edwards
   (1989), Genders (1963a), Jelitto & Schacht (1990), Ogg (1961), and the Royal
   Horticultural Society (1926e) also list many cultivars. Kelsey & Dayton (1942)
   also lists delphiniums but without introduction dates and background.

Dendranthema The cultivars of *D.* x *grandiflorum* (Ramat.) Kitamura, or florist's
   chrysanthemum, are first listed by Olmsted et al. (1923) and Kelsey & Dayton
   (1942). These lists omit all cultivars 1789–1889. Cumming (1964) presents some
   history. Scott & Scott (1950) are also recommended for early cultivars and
   hybridizers. Additional cultivars are listed by Gosling (1964, 1973b, 1980) and
   the [U.S.] National Chrysanthemum Society (1991). Genders (1961), Gosling
   (1973a), Jelitto & Schacht (1990), and Trehane (1989) also list some cultivars.
   Niwa (1936) should be consulted for Japanese cultivars.

Desmodium The cultivars of the tick trefoils are listed by Lemmens (1985).

Deutzia The cultivars of the deutzias are discussed by Bean (1970–1988) and
   Krüssmann (1958b).

Dianthus The Royal Horticultural Society has published the international register for
   pinks and carnations (1983) with supplements (1984b, 1985b, 1986b, 1988b,
   1988c, 1989b, 1990a), which supersede the 1974 list. These lists are very
   comprehensive, but American cultivars (such as 'Aqua') are slow to be
   integrated. These checklists should be used in conjunction with Bailey (1990),
   Mansfield (1951), Sitch (1975), and Smith (1990). The history and performance
   of *D. gratianopolitanus* Vill. and *D. plumarius* L. are discussed in Hensen
   (1981). Kelsey & Dayton (1942) include American carnation cultivars. Jelitto &
   Schacht (1990) and Trehane (1989) also list many cultivars of perennial *Dianthus*.

Diastema See *Gesneriaceae*.

Diervilla The cultivars of the bush honeysuckles are discussed by Schneider (1930).

Dracaena The Indian cultivars of the dracaenas are listed by Anonymous (1959d).

Drosera See carnivorous plants.

Echeveria Carruthers & Ginns (1973) list cultivars of *Echeveria* but provide no dates.

Epacris See Australian & South African plants.

Epimedium The cultivars of the epimediums are listed by Laar (1981a).

Epiphyllum See *Cactaceae*.

Episcia See *Gesneriaceae*.

Eremophila See Australian & South African plants.

Erica A guide to naming heath cultivars is McClintock (1986). Bean (1970–1988),
   Chapple (1951), Johnson (1956), Krüssmann (1984–1986), Laar (1970a, 1977a),
   Letts (1966), Proudley & Proudley (1974), Underhill (1990), and Laar (1974) list
   many cultivars of heaths. Munson (1981, 1984) provides a comprehensive key
   to species and cultivars of *Erica* with full botanical descriptions in the 1981
   thesis.

Erigeron The history and performance of cultivars of the daisy fleabanes are discussed
   by Clausen (1972b), Hensen (1966), Jelitto & Schacht (1990), and Oudshoorn
   (1975). Trehane (1989) also lists many cultivars of the daisy fleabanes.

Eriostemon See Australian & South African plants.

Erodium The cultivars of the heron's bills are listed by the British Pelargonium and
   Geranium Society (1970).

Erythrina The species and cultivars of *Erythrina* are listed by McClintock (1982).

Escallonia The cultivars of the escallonias are listed by Bean (1970–1988), Krüssmann (1984–1986), and Schneider & Laar (1970).
Eucalyptus See Australian & South African plants.
Eucodonia See *Gesneriaceae*.
Eucryphia The cultivars of *Eucryphia* are discussed by Wright (1983a) but with no introduction dates or background. Bean (1970–1988) and Krüssmann (1984–1986) also list some cultivars. Recent registrations are listed by the Australian Cultivar Registration Authority (1988).
Euonymus The cultivars of *Euonymus* are discussed by Bean (1970–1988), Ilsink & Jong (1986), Krüssmann (1984–1986), Laar (1979), and Lancaster (1981). An additional registration is recorded by Huttleston (1986).
Euphorbia Ecke (1976) lists cultivars of the poinsettia, *Euphorbia pulcherrima* Willd. ex Klotzsch. Additional registrations are recorded by Huttleston (1989). Trehane (1989) and Turner (1983) discuss the cultivars of the hardy spurges.
Fagus Wyman (1962a, 1964) lists the cultivars of the beeches but with few introduction dates and little background. Bean (1970–1988), Grootendorst (1975), and Krüssmann (1984–1986) also discuss the cultivars of the beeches. Spongberg (1988, 1989) records additional registrations.
Ferns Maatsch (1980) lists cultivars of ferns with descriptions, background information, and English vernacular names. Jones (1987) and Mickel (1994) list many cultivars of ferns. The cultivars of *Adiantum,* maidenhair ferns, are listed in Goudey (1985). The cultivars of the staghorn ferns, *Platycerium,* are listed by Vail (1984). The cultivars of the royal ferns, *Osmunda regalis* L., are listed by Anderson (1971). Recent registrations of *Blechnum* are listed by the Australian Cultivar Registration Authority (1988).
Forsythia The cultivars of the forsythias are published by Bean (1970–1988), Hebb (1971), Krüssmann (1984–1986), Werken (1988), and Wyman (1961a, 1961b). Spongberg (1988) records additional registrations.
Fraxinus The cultivars of the ashes are discussed by Bean (1970–1988), Bom (1982), Grootendorst (1966b), Krüssmann (1984–1986), McArdle & Santamour (1984), and Santamour & McArdle (1983c).
Fuchsia Parker (1986) provides a guide for fuchsia cultivar identification. A comprehensive checklist of cultivars of fuchsias is Boullemier (1975, 1980, 1982, 1985). Bean (1970–1988) and Krüssmann (1984–1986) discuss some hardy cultivars. Ewart (1982, 1987) and Saunders (1971–1973) describe and picture many of these cultivars. These should be used in conjunction with Manthey (1990), Proudley (1975), and Thorne (1959).
Gaillardia The cultivars and performance of the gaillardias are listed by the Royal Horticultural Society (1930c).
Galanthus The cultivars of the snowdrops are listed by Bowles (1956), the Royal General Bulbgrowers' Association (1991), Trehane (1989), and Yeo (1975).
Genista The cultivars of the brooms *(Cytisus* and *Genista)* are treated by Bean (1970–1988), Krüssmann (1984–1986), and Laar (1971).
Gentiana Bartlett (1975) includes many cultivars of gentians in her book. Many cultivars are also listed by Trehane (1989).
Geranium The history and performance of cultivars of the hardy geraniums are discussed by Clausen (1974b). Yeo (1985) includes information on many cultivars of the hardy geraniums in his comprehensive book. Clifton (1992), Jelitto & Schacht (1990), Trehane (1989), and Walsweer (1988) list many cultivars.
Gesneriaceae Cultivars of *Achimenes,* the orchid pansies, are listed by the American Gesneria Society (1957), Arnold (1969), and Townsend (1984). Cultivars of *Aeschynanthus* are listed by Dates (1990). Cultivars of *Columnea* and allied

genera (*Bucinellina, Dalbergaria, Pentadenia,* and *Trichantha*) are listed by Arnold (1963b, 1966) and Dates (1987). Cultivars of *Episcia* and *Alsobia,* the carpet plants, are listed by the American Gesneria Society (1957), Arnold (1963a, 1968, 1977), and Dates (1993). Cultivars of *Kohleria,* the tree gloxinias, and *Smithiantha* are listed by the American Gesneria Society (1957), the American Gloxinia Society (1962), Batcheller (1985), and Moore (1953). Cultivars of *Nematanthus* are listed by Arnold (1978). The master variety lists for *Saintpaulia,* the African violets, are published by Boland (1983, 1984, 1985, 1986, 1987, 1988), Frank (1975), Rector (1963), and Tretter (1976) and should be supplemented with Kawakami (c. 1981), who includes English captions and a Japanese text with color photographs. Cultivars of *Sinningia* are listed by Arnold (1975) and Dates (1988). Cultivars of *Streptocarpus,* the Cape primroses, are listed by Arnold (1972, 1979) and Brown (1973). Intergeneric hybrids in the tribe Gloxinieae (which includes *Achimenes, Diastema, Eucodonia, Gloxinia, Heppiella, Koellikeria, Kohleria, Monopyle, Moussonia, Niphaea, Parakohleria, Smithiantha,* and *Solenophora*) are listed by Dates (1986).

Geum The history and performance of cultivars of *Geum* are discussed by Clausen (1975) and Mallett (1983). Jelitto & Schacht (1990) and Trehane (1989) also list many cultivars.

Ginkgo The cultivars of ginkgo (*G. biloba* L.) are discussed by Bean (1970–1988), Bom (1982), Krüssmann (1984–1986), and Santamour, He, & McArdle (1983).

Gladiolus Pieters (1905) is the first list of cultivars of gladiolus that we have discovered. The cultivars of gladiolus are later listed by Hottes (1916), Olmsted et al. (1923), the American Gladiolus Society (1931), Birch (1940), Pridham (1932), the Royal General Bulbgrowers' Association (1991), Sandhack (1927), and Trehane (1989). Many of these cultivars are incorporated into Fisher et al. (1975). Additions are published by Fisher (c. 1983).

Gleditsia Bean (1970–1988), Bom (1982), Haserodt & Sydnor (1983), Krüssmann (1984–1986), Santamour & McArdle (1983b), and Wagenknecht (1961a) discuss the cultivars of the honey locust (*G. triacanthos* L.).

Gloriosa Narain (1988), the Royal General Bulbgrowers' Association (1991), and Trehane (1989) list cultivars of the climbing lilies.

Gloxinia See *Gesneriaceae.*

Grevillea The grevilleas are discussed by Larkman (1985). Recent registrations are listed by the Australian Cultivar Registration Authority (1988). See also the entry for *Proteaceae.*

Hakea See Australian & South African plants and *Proteaceae.*

Halesia The cultivars of the silverbells are listed by Fontaine (1970b).

Hamamelidaceae The cultivars of the *Hamamelidaceae* are discussed by Bean (1970–1988), Krüssmann (1984–1986), Sanders (1982), Weaver (1976a), and Wright (1982). Grootendorst (1965, 1980a) and Lancaster (1970) record the background and performance of witch hazel (*Hamamelis* spp.) cultivars. Huttleston (1989) records an additional registration of *Hamamelis.*

Hardenbergia See Australian & South African plants.

Hebe Chalk (1988) lists cultivars of *Hebe* and *Parahebe.* Bean (1970–1988) and Krüssmann (1984–1986) also discuss the cultivars of these two genera.

Hedera The most comprehensive listing of cultivars of ivies is Heieck (1980). This should be supplemented with Fearnley–Whittingstall (1992), Hatch (1982), Krüssmann (1984–1986), Lawrence & Schulze (1942), Lawrence (1956), Nannenga–Bremekamp (1970), Pierot (1974), Rose (1980), Schaepman (1975), and publications in the *Ivy Journal.*

Hedychium The cultivars of the ginger lilies are discussed by Schilling (1982).

Hedysarum The cultivars of the *Hedysarum* species are listed by Lemmens (1985).
Helianthemum The history and performance of cultivars of the rock roses are
   discussed by Clausen (1968) and the Royal Horticultural Society (1926d). Jelitto
   & Schacht (1990) and Trehane (1989) also list cultivars.
Helianthus The history and performance of cultivars of the sunflowers are discussed
   by Clausen (1974c). Trehane (1989) also lists cultivars.
Helichrysum See Australian & South African plants.
Heliconia The cultivars of *Heliconia* are discussed by Berry & Kress (1991).
Heliopsis The history and performance of cultivars of *Heliopsis* are discussed by
   Clausen (1974a) and Hensen (1983b). Jelitto & Schacht (1990) and Trehane
   (1989) also list cultivars.
Helleborus The cultivars of the hellebores are discussed and illustrated in black and
   white drawings by Ahlburg (1993) and appended to the rear of Mathew (1989b),
   but the most comprehensive descriptions, with color photographs, seem to be
   Rice & Strangman (1993).
Hemerocallis The daylilies are first listed by Stout (1934) and Norton et al. (1949),
   and later in publications of the American Hemerocallis Society (1957, c. 1973,
   c. 1984). The species and old cultivars are discussed by Kitchingham (1985).
   Jelitto & Schacht (1990) and Trehane (1989) also list cultivars. Additional
   information is provided by Munson (1989) and Webber (1988). These should be
   used in conjunction with Darrow & Meyer (1968), Erhardt (1992), and Stout
   (1986).
Heppiella See *Gesneriaceae*.
Heuchera The history and performance of the coralbells are discussed by Clausen
   (1970) and Hansen & Sieber (1970). Jelitto & Schacht (1990) and Trehane (1989)
   also list cultivars.
Hibiscus Krüssmann (1984–1986) discusses the cultivars of *Hibiscus*. The American
   Hibiscus Society (1984, 1987) and Chin (1986) are checklists and illustrated
   catalogs of cultivars of Chinese hibiscus (*H. rosa-sinensis* L.). Beers & Howie
   (1985, 1990), Harvey (1988), and Howie (1980) are checklists of mostly
   Australian cultivars of Chinese hibiscus. Cultivars of rose of Sharon (*H. syriacus*
   L.) are published by Bean (1970–1988), Grootendorst (1968b), Huttleston (1986,
   1988, 1990, 1991), and Wyman (1958). Kelsey & Dayton (1942) provide a list
   of cultivars for both species but without introduction dates and background.
Hippeastrum See *Amaryllidaceae*.
Hoheria The cultivars of the lacebarks are listed by Bean (1970–1988), Krüssmann
   (1984–1986), and Wright (1983a).
Hosta The most comprehensive discussion of the hostas is Schmid (1991). The
   cultivars of the hostas are also listed in the comprehensive works by Hensen
   (1963a, 1963b, 1983a,1985) with comments by Grenfell (1986). These should be
   supplemented with Aden (1990), Fisher (1979), Grenfell (1990), Jelitto & Schacht
   (1990), Laar (1967), and Trehane (1989).
Houseplants Cultivars of houseplants are pictured and briefly discussed by Graf
   (1986a, 1986b).
Hyacinthus The cultivars of hyacinths are documented by the Royal General
   Bulbgrowers' Association (1991), but further information is provided by
   Darlington, Hair, & Hurcombe (1951). Trehane (1989) also lists the cultivars of
   hyacinths.
Hydrangea The most comprehensive discussion of the cultivars of the hydrangeas is
   Mallet, Mallet, & van Trier (1992). Cultivars of the hydrangeas are also covered
   in Bean (1970–1988), Grootendorst (1973b), Krüssmann (1984–1986), and

Haworth–Booth (1984). Ilsink (1988) covers the cultivars of *H. paniculata* Sieb., while Wilson (1923) covers the cultivars of *H. macrophylla* (Thunb.) Ser.

Hypericum The cultivars of the St. Johnsworts are listed by Bean (1970–1988) Krüssmann (1984–1986), and Schneider (1965b, 1966a).

Hypocalymma See Australian & South African plants.

Ilex In 1953 the Holly Society of America published a preliminary checklist (Wister, 1953b). This checklist is being revised in accordance with the International Code of Nomenclature for Cultivated Plants. First in this series is Eisenbeiss & Dudley (1973) for *I. opaca* Aiton, the American holly. Eisenbeiss & Dudley (1983) is an *Ilex* cultivar registration list 1958–1983. Andrews (1983, 1984, 1986) and Gelderen (1988) discuss the cultivars of *I.* x *altaclerensis*. The varieties of *I. aquifolium* L. are listed in Gelderen (1988) and Paul (1863). Wyman (1960) and Dudley & Eisenbeiss (1992) are checklists for *I. crenata* Thunb. Dirr (1988) and Eggerss & Hasselkus (1992) cover the cultivars of the deciduous hollies. Bean (1970–1988), Gelderen (1971), and Krüssmann (1984–1986) also list the cultivars of the hollies. These should be used in conjunction with Hansell, Dudley, & Eisenbeiss (1970). New cultivars of hollies are published in the *Holly Society Journal*.

Impatiens The cultivars of the New Guinea impatiens are discussed by Agnew & Lang (1992), Eichin & Deiser (1988) and Winters (1973).

Indigofera The cultivars of the indigos are listed by Lemmens (1985).

Iris Jelitto & Schacht (1990) and Trehane (1989) list many cultivars of the hardy iris but with scant information. In contrast, the listings of cultivars of bearded irises published by Peckham (1929, 1940), Douglas (1949), Knowlton (1959), Nelson (1971), and Nelson & Keppel (1981, 1991, 1992a, 1992b, 1993) are rich with information. The Royal Horticultural Society (1928, 1930b) and Sand (1925) provide additional descriptions of many pre–1930 bearded irises. While the American Iris Society has published yearly checklists since the 1979 checklist, the 1989 checklist is still in press.

Cultivars of the reblooming iris are listed by Brookins (1991a). Cultivars of the dwarf irises are listed by the Dwarf Iris Society (1975, 1988) and Wright (1927). Cultivars of Japanese iris (*I. kaempferi* Sieb. ex Lem.) are published by Brookins (1992) and the Society for Japanese Irises (1988), supplemented by McEwen (1990). Cultivars of the Siberian irises (*I. sanguinea* Hornem. and *I. sibirica* L.) are listed by Brookins (1991b) and Warburton (1986). Cultivars of the arils are listed by the Aril Society (1976, 1978, 1980, 1982). The Louisiana irises are listed by Fritchie (1982), and some color photographs, dates, and background are given by Caillet & Metzweiller (1988). Cultivars of the medians are listed by the Median Iris Society (1984, 1992). Cultivars of the spurias are listed by Foreman (1985). Cultivars of the bulbous irises are discussed by Hoog (1980) and the Royal General Bulbgrowers' Association (1991).

Ixora Anonymous (1958c) is a checklist of the cultivars of the ixoras.

Jasminum The cultivars of the jasmines are discussed by Bean (1970–1988), Green (1965), and Krüssmann (1984–1986). An additional cultivar is listed by Huttleston (1986).

Juniperus See conifers.

Kalmia The cultivars of the mountain laurels are published in Jaynes (1975, 1983, 1988) and Krüssmann (1984–1986); additional cultivars are published in *HortScience* (Jaynes, 1989).

Kennedia See Australian & South African plants.

Kniphofia The cultivars of the torch lilies are discussed by Jelitto & Schacht (1990), Taylor (1985a, b), and Trehane (1989).

Koellikeria See *Gesneriaceae*.
Kohleria See *Gesneriaceae*.
Kunzia See Australian & South African plants.
Lagerstroemia The cultivars of crape myrtles are listed by Egolf & Andrick (1978) and Krüssmann (1984–1986).
Lantana The cultivars of the lantanas are listed by Anonymous (1958d), Howard (1969), and Krüssmann (1984–1986). An additional cultivar is recorded by Spongberg (1988).
Lathyrus Kelsey & Dayton (1942), Royal Horticultural Society (1926b), and Unwin (1926) are checklists of sweet peas, *Lathyrus odoratus* L. Cultivars of the hardy perennial species are listed by Trehane (1989).
Lavandula The cultivars of lavender (*L. angustifolia* Mill.) and lavandin (*L.* x *intermedia* Emeric ex Loisel.) are discussed by Hensen (1974), Krüssmann (1984–1986), and Tucker & Hensen (1985).
Lechenaultia See Australian & South African plants.
Leptospermum Krüssmann (1984–1986) and Metcalf (1963) are checklists of *Leptospermum* cultivars (mostly *L. scoparium* J. R. Forst & G. Forst). Recent registrations are listed by the Australian Cultivar Registration Authority (1988).
Lespedeza The cultivars of the bush clovers are listed by Lemmens (1985). An additional cultivar is listed by Huttleston (1991).
Leucadendron See *Proteaceae*.
Leucospermum See *Proteaceae*.
Leucothoè The cultivars of *L. fontanesiana* (Steud.) Sleum., the drooping leucothoè, are discussed by Bean (1970–1988), Krüssmann (1984–1986), and Green (1963).
Lewisia The cultivars of the lewisias are discussed by Mathew (1989a).
Ligularia The cultivars of *Ligularia* are discussed by Dress (1962).
Lilium Checklists of lilies are published by Leslie (1982) with supplements (Royal Horticultural Society, 1982, 1984a, 1985a, 1986a, 1987a, 1988a, 1989a, 1990b, 1991a, 1992b, 1993a, 1994). The North American Lily Society also has its checklists of lily hybrids (Fisher, 1978; Collings, 1986) but gives scant information on cultivars pre–1940. Additional listings are carried by Trehane (1989).
Liquidambar The cultivars of the the the sweet gums (*L. formosana* Hance and *L. styraciflua* L.) are discussed by Bom (1982), Krüssmann (1984–1986), and Santamour & McArdle (1984).
Liriodendron The cultivars of the tulip tree (*L. tulipifera* L.) are discussed by Bean (1970–1988), Krüssmann (1984–1986), and Santamour & McArdle (1984).
Liriope See *Ophiopogon–Liriope*.
Lobelia The tetraploid cultivars of the *Lobelia siphilitica–L. cardinalis* complex are listed by Bowden (1983).
Lonicera The cultivars of the honeysuckles are discussed by Bean (1970–1988), Krüssmann (1984–1986), Laar (1988), Schneider (1971), Wright (1983b), and Yeo (1964).
Lophostemon See Australian & South African plants.
Lupinus The cultivars and performance of the lupines are discussed by the Royal Horticultural Society (1931a).
Lythrum The cultivars of *Lythrum* are discussed by Harp (1975).
Magnolia Tresender (1978) lists many cultivars of magnolias, and while dates and hybridizers are generally absent, the descriptions are good. Gardiner (1989) thoroughly discusses magnolia hybrids with photographs. Additional registrations are listed in Bean (1970–1988), Krüssmann (1984–1986), and Vrugtman (1972).

Fogg & McDaniel (1975) is a comprehensive list of magnolia cultivars. New cultivars of magnolias are published in *Magnolia Journal.*

Mahonia The cultivars of the grape hollies (and x *Mahoberberis)* are listed by Bean (1970–1988), Brickell (1979), Krüssmann (1984–1986), and Laar (1975).

Malus Bom (1982), den Boer (1959), Grootendorst (1964a), Lombarts (1984), Preston (1944), Van Eseltine (1933, 1934), and Wyman (1943, 1955) include descriptions and introductions of crabapples (the 1943 edition of Wyman has some information dropped from the 1955 edition, including discarded cultivars and citations to a bibliography). Jefferson (1970) clarifies the misnaming of crabapple cultivars and provides an extensive bibliography. Bean (1970–1988) and Krüssmann (1984–1986) also list the cultivars of the ornamental crabapples. Lately, crabapple registrations have been carried by the Arnold Arboretum and published in *HortScience* (Spongberg, 1988, 1989).

Melaleuca See Australian & South African plants.

Melia The cultivars of the Persian lilac or Chinaberry (*Melia azedarach* L.) are listed by Mabberly (1984).

Monarda The beebalms are discussed by Oudolf (1993).

Monopyle See *Gesneriaceae.*

Moussonia See *Gesneriaceae.*

Myoporum See Australian & South African plants.

Nandina The landscape values of cultivars of heavenly bamboo, *N. domestica* Thunb., are discussed by Raulston (1984).

Narcissus For many years the Royal Horticultural Society printed classified lists of daffodils (1908d, 1910, 1931b, 1938, 1948, 1955, 1958, 1961, 1965, 1969c, 1975), but many of these were not cumulative. The most comprehensive list of cultivars of daffodils is by the Royal Horticultural Society (Kington, 1989a), which updates the 1969 classified list and the classified list and international register of 1975 with supplements 1–14; supplements 15–18 are printed separately (Kington, 1989b, 1990, 1991, 1992). A checklist of daffodils has been provided as a continually updated computer printout by Throckmorton (n.d.), but very old cultivars are listed without a date. The history of cultivar registration of daffodils by the Royal Horticultural Society is documented by Donald (1986). Trehane (1989) also lists cultivars of daffodils. Abridged lists of exhibition daffodils are published by the American Daffodil Society (1977, 1985, 1989). These should be used in conjunction with Bourne (1903), Bowles (1934), Lee (1966), Tompsett (1982), and the "Narcissus editions" of *Herbertia* (vol. 13, 1946) and *Plant Life/Herbertia* (vol. 9, no. 1, 1953).

Nematanthus See *Gesneriaceae.*

Nepenthes See carnivorous plants.

Nerine See *Amaryllidaceae.*

Nerium The cultivars of the oleanders are discussed by Anonymous (1958a) and Pagen (1987).

Nigella The cultivars of *Nigella* are listed by Sorvig (1983).

Niphaea See *Gesneriaceae.*

Nymphaeaceae The most comprehensive list of waterlily cultivars is Swindells (1989b). The cultivars of waterlilies are also treated in Anonymous (1960); Conard (1905); Henkel, Rehnelt, and Dittmann (1907); Kelsey & Dayton (1942); Swindells (1983); and Trehane (1989).

Ophiopogon-Liriope The cultivars of *Ophiopogon* and *Liriope* are discussed by Hume (1961).

Orchidaceae A guide to orchid hybrid (grex) registration is published by Hunt (1986). While lists have been previously published by Sanders, Sanders (1946) is the last cumulative checklist of orchid hybrids; a list of intergeneric taxa is listed in Table II. Later, noncumulative supplements have been published (Sanders & Wreford, 1961; Royal Horticultural Society, 1972, 1980, 1981, 1985c, 1986c). Japan Orchid Growers Association (n.d.) has excellent color photographs of cultivars derived from *Cattleya*. Only Poliakoff (1987) lists *Vanda* cultivars with the percentage of genetic background of each ancestral species. Gilmour, Greatwood, and Hunt (1976) give the names of intergeneric hybrids.

Origanum The cultivars of *Origanum,* the marjorams, are discussed by Tucker & Rollins (1989). Trehane (1989) lists some additional cultivars.

Osmunda See ferns.

Ostrya See *Carpinus.*

Paeonia The cultivars of the peonies are first listed by Coit (1907), later by Beal (1920) and Kelsey & Dayton (1942), and most recently by Jelitto & Schacht (1990) and Trehane (1989). The most comprehensive listing is by Kessenich (1976). These checklists should be supplemented with Wister (1962) for fuller descriptions and a comprehensive bibliography. Haworth–Booth (1963) and Krüssmann (1984–1986) also supply further information on the tree peonies. American peony hybrids are listed by Kessenich (1990). Later introductions have been published in the *American Peony Society Bulletin.*

Pandorea See Australian & South African plants.

Papaver The primary reference on poppy cultivars is Grey–Wilson (1993). The cultivars of the oriental (*P. orientale* L.), Iceland (*P. nudicaule* L.), and other poppies are also listed by Kelsey & Dayton (1942).

Parakohleria See *Gesneriaceae.*

Parthenocissus The cultivars of Boston ivy, *P. tricuspidata* (Sieb. & Zucc.) Planch, are listed by Laar (1981b, 1992).

Passiflora The cultivars of *Passiflora,* the passion flowers, are thoroughly discussed by Vanderplank (1991).

Pelargonium The most readily available guide to geranium cultivars is Krauss (1955). Bagust (1988) lists the cultivars of the dwarf geraniums. Moore (1955a, 1955b) provides background information on many species and some cultivars. The Australian Geranium Society (1978, 1985) has published the first two sections of a comprehensive *Pelargonium* checklist. Clifford (1970) is also useful.

Penstemon Lindgren (1993) has provided a guide to registration of *Penstemon* cultivars. The most comprehensive guide to the genus is Lindgren & Davenport (1992). The American Penstemon Society (McWilliam, 1973, 1977) also lists registered cultivars.

Pentadenia See *Gesneriaceae.*

Perennials, herbaceous Grunert (1982), Jelitto & Schacht (1990), Krüssmann, Siebler, & Tangermann (1970), Phillips & Rix (1991), Thomas (1990), and Wehrhahn (1931) rank high among the available reference works on hardy herbaceous plants because of the wealth of information. The perennials registered by the International Registration Authority for Hardy Perennial Plants are listed by Sieber (1990a, 1990b). The cultivars of perennials, based primarily upon British catalogs, are listed by Philip (1992); perennial cultivars based upon northern European sources are listed by Laar & Fortgens (1990). Cultivars of perennials based upon American catalogs are listed by Isaacson (1989). Trehane (1989), emphasizing the cultivars available in the United Kingdom and Northern Europe, provides many dates and names of introducers. Though these latter three publications are excellent, they reinforce some incorrect synonyms by uncritically accepting catalog listings.

Pernettya The cultivars of *Pernettya* are listed by Laar (1969) and Vogel (1969).

Petunia *Petunia* cultivars are listed by Maatsch & Nolting (1968, 1971b) and Nolting & Zimmer (1975b, 1980b, 1984, 1987); the earliest cultivar in these is dated 1947.

Philadelphus The mock orange cultivars are listed by Bean (1970–1988), Dolatowski (1986), Hu (1954–1956), Janaki Ammal (1951), Kapranova & Lukina (1972), Krüssmann (1958c, 1984–1986), Sampson (1965), Schneider (1934), Wright (1980), and Wyman (1965). Recent mock orange cultivars are published by Huttleston (1988).

Phlox Probably the most comprehensive list of phlox culivars is Trehane (1989), but very few dates are provided. Jelitto & Schacht (1990), Kelsey & Dayton (1942), Kharchenko (1975), and Symons-Jeune (1953) also list phlox cultivars.

Phormium The cultivars of New Zealand flax are discussed by Cheek (1979) but more thoroughly by Heenan (1991). New cultivars are listed by Hornback (1994).

Phygelius Cultivars of *Phygelius* are discussed by Coombes (1988). Trehane (1989) also lists cultivars.

Picea See conifers.

Pieris The cultivars of the Japanese andromedas are listed by Bean (1970–1988), Bond (1982), Gelderen (1979), Ingram (1963), Krüssmann (1984–1986), and Wagenknecht (1961b). Spongberg (1988, 1990) records additional registrations.

Pimelea See Australian & South African plants.

Pinus See conifers.

Plagianthus The cultivars of *Plagianthus* are listed by Wright (1983a).

Plant patents The U.S. plant patents and their common names have been assembled by the American Association of Nurserymen (1957, 1958, 1959, 1960, 1961, 1962, 1963, 1967, 1969, 1974, 1981) for plant patents 1–4359. These have been published in one directory (American Association of Nurserymen, 1990) with patents 1–7088. Patents 1–477 are also listed in Kelsey & Dayton (1942).

A review of the U.K. system of Plant Breeders' Rights (PBR) is Goodwin (1986). The patenting of plants under the European Patent Convention (EPC) and The International Union of the Protection of New Varieties of Plants (UPOV) has been reviewed by Byrne (1986), Mast (1986), and Schneider (1986b).

Platanus The cultivars of the plane trees are discussed by Santamour & McArdle (1986).

Platycerium See ferns.

Plectranthus See *Coleus-Plectranthus.*

Plumeria The checklist of the Plumeria Society of America (1988) should be supplemented with the color photographs of Chinn & Criley (1982), Eggenberger & Eggenberger (1988), and Thornton & Thornton (1985). Another checklist of cultivars is Anonymous (1958b).

Poaceae, Cyperaceae, and Juncaceae The best listings of the ornamental grasses, sedges, and rushes are Darke (1990), Hensen & Groendijk-Wilders (1986b), and Trehane (1989). These should be supplemented with Jelitto & Schacht (1990), Loewer (1988), Meyer (1975), Grounds (1979), Ottesen (1989), and Reinhardt et al. (1989). Lawson (1968) lists some cultivars of bamboos.

Populus Checklists of poplar cultivars are Broekhuizen (1977), International Poplar Commission (1971, 1990), Koster (1972), and Roller, Thibault, & Hidahl (1972). Bean (1970–1988) and Krüssmann (1984–1986) provide additional information.

Potentilla Hachmann et al. (1986a), Jelitto & Schacht (1990), Schmalscheidt (1984), and Trehane (1989) list the cultivars of *Potentilla*. The cultivars of the shrubby potentillas are discussed by Bachtell & Hasselkus (1982), Bean (1970–1988), Brearley (1987), Krüssmann (1984–1986), and Rhodes (1954). The cultivars of

    P. fruticosa L. are listed by Bowden (1957), Laar (1982), Schneider (1967), and Wyman (1968). An additional registration is discussed by Huttleston (1990).

Primula The best listings of cultivars of *Primula* is Trehane (1989). Blasdale (1948), Genders (1962, 1963b), Haysom (1957), Hecker (1971), Hyatt (1989), Jelitto & Schacht (1990), Lyall (1959), Puttock (1957), Swindells (1989a), and Wemyss-Cooke (n.d.) also list many cultivars.

Prostanthera The few cultivars of the mint shrubs are briefly mentioned by Althofer (1978).

Protea See *Proteaceae*.

Proteaceae Matthews (1983, 1993) and Vogts (1982) provide descriptions and excellent colored illustrations of cultivars of genera of the *Proteaceae*: *Banksia, Grevillea, Hakea, Leucadendron, Leucospermum, Protea, Serruria,* and *Telopea*. A guide to cultivar registration for *Proteaceae* is presented by Brits (1988a, c), while a "sample list" of *Proteaceae* cultivars is presented by Brits (1988b).

Prunus Chadbund (1972) is recommended for cultivars of many flowering cherries. The Oriental flowering cherries are listed by Russell (1934), while the purpleleaf plums are discussed by Jacobson (1992). Only the Sato-zakura group of the Japanese flowering cherries has been published as a separate checklist (Jefferson & Wain, 1984). The bibliography of this checklist, however, gives invaluable references on other ornamental *Prunus*. These should be supplemented with Bom (1982), Grootendorst (1964b), Ingram (1948), Laar (1970b), Miyoshi (1916), Ohwi & Ohta (1973), and Wilson (1916). Other ornamental *Prunus* are listed by Bean (1970–1988), Huttleston (1986, 1990), and Krüssmann (1984–1986).

Pterostyrax The cultivars of the epaulette trees are listed by Fontaine (1970b).

Pulmonaria Cultivars of the lungworts are reviewed by Mathew (1982), Jelitto & Schacht (1990), and Trehane (1989) but generally without introduction dates or names of originators.

Pultenaea See Australian & South African plants.

Pyracantha Cultivars of the fire thorns are listed by Bean (1970–1988), Laar (1966), Hachmann et al. (1986b), Krüssmann (1984–1986), Schmalscheidt (1984), and de Vos (1958).

Pyrus Bean (1970–1988), Bom (1982), and Krüssmann (1984–1986) list ornamental cultivars of pears. The cultivars of the Callery pear (*P. calleryana* Decne.) are discussed by Santamour & McArdle (1983a).

Quercus The cultivars of the oaks are listed by Wyman (1962f) but with few introduction dates and little background. Bean (1970–1988), Bom (1982), Grootendorst (1980b), Krüssmann (1984–1986), and McArdle & Santamour (1985, 1987a, 1987b) thoroughly discuss the cultivars of oaks.

Rhododendron Brickell (1980) provides guidelines for naming *Rhododendron* cultivars. The cultivars of rhododendrons and azaleas are first documented in Fletcher (1958), and this is updated by Royal Horticultural Society (1964, 1969b, 1988e, 1989d, 1989e, 1990c, 1991b, 1992c, 1993b); the registrations from 1962 to 1987 were originally published in *The Rhododendron and Camellia Yearbook* and *Rhododendron with Magnolias and Camellias*. Kraxberger (1980) lists American *Rhododendron* hybrids, many of which were originally published in *Rhododendrons* and *Rhododendron Notebook;* more recently the American hybrids have been listed in the *Journal of the American Rhododendron Society*. German *Rhododendron* hybrids are discussed by Schmalscheidt (1980). These checklists should be used in conjunction with Bean (1970–1988), Bowers (1960), Bulgin (1986), Cox (1985), Cox & Cox (1988), Galle (1985), Gelderen & Hoey Smith (1992), Greer (1982), Grootendorst (1954, 1967b, 1969b, 1969c, 1979a), Ihei (1984), Krüssmann (1984–1986), Leach (1961), Lee et al. (1965), Livingston &

West (1978), Morrison (1953), Phillips & Barber (1967, 1979), Salley & Greer (1986, 1992), and Schneider (1965c, 1966b).

Robinia The cultivars of the locusts are listed by Bean (1970–1988), Gibbs (1929), Grootendorst (1971a), and Krüssmann (1984–1986).

Rosa A proposed guide to rose name registration is Gioia (1986). The most comprehensive recent checklist of cultivars of roses is *Modern Roses 10* (Cairns, 1993), but *Modern Roses 9* (Haring, 1986), *Modern Roses 8* (Meikle, 1980) and *Modern Roses 6* (Allan, 1965) are also important for some rose cultivars. Stock (1984) lists the older and foreign checklists (especially important for heritage roses), such as Boitard (1836), Desportes (1828), Gravereaux (1902), Mansfield (1943), Nietner (1880), Park (1956), Simon & Cochet (1906), and Singer (1885). Jäger (1960) is a reprint of a privately distributed list of 1936. The Royal National Rose Society also publishes selected checklists, most recently in 1976. These should be used in conjunction with Austin (1988), Beales (1985, 1988), Bean (1970–1988), Dickerson (19932), Fagan (1988), Gault & Singe (1971), Griffiths (1984, 1987), Harkness (1991), Krüssmann (1981), Moody (1992), and Phillips & Rix (1988). The *Combined Rose List* (Dobson, 1987, 1988, 1989, 1990, 1991; Dobson & Schneider, 1992) provides continued updating of available roses around the world with cultivar information.

Rosmarinus The origins and essential oils of cultivars of rosemary are listed by Tucker & Maciarello (1986).

Saintpaulia See *Gesneriaceae.*

Salix Newsholme (1992) provides the most comprehensive descriptions of cultivars of *Salix,* the willows. The cultivars of the weeping willow (*S. babylonica* L.) are discussed by Santamour & McArdle (1988). Broekhuizen & Schneider (1969) discusses the cultivars of the white willow (*S. alba* L.). Anonymous (n.d.), Bean (1970–1988), and Krüssmann (1984–1986) also discuss cultivars of willows.

Sambucus The cultivars of European red elderberry, *S. racemosa* L., are described in German and Latin by Wolf (1923). Bean (1970–1988) and Krüssmann (1984–1986) also discuss cultivars of the elderberries.

Sansevieria The cultivars of the snakeplants are listed by Morgenstern (1979), Stover (1983), and Swinbourne (1979) but without introduction dates and background. Chahinian (1986) thoroughly treats the cultivars of *S. trifasciata.*

Sarracenia See carnivorous plants.

Saxifraga The saxifrages are listed by Köhlein (1984) but without introduction dates or background. Jelitto & Schacht (1990), Trehane (1989), and Webb & Gornall (1989) are comprehensive lists of species and cultivars.

Scabiosa The annual derivatives of *S. atropurpurea* L. are listed by the Royal Horticultural Society (1926c). Perennial cultivars are listed by Jelitto & Schacht (1990).

Scaevola See Australian & South African plants.

Schizostylis The cultivars of the Kaffir lily are listed by Straley (1984).

Schlumbergera See *Cactaceae.*

Sedum Praeger (1921) and Trehane (1989) list the cultivars of *Sedum.* The history and performance of cultivars of *Sedum* are discussed by Clausen (1978). Hensen & Groendijk-Wilders (1986a) discuss the sedums cultivated in Europe. Some cultivars of sedums are listed by Evans (1983) and Jelitto & Schacht (1990) but without introduction dates or background.

Sempervivum The cultivars of *Sempervivum* (and *Jovibara*) are listed by Mitchell (c. 1973) with some color photographs and good descriptions but without dates or background. Subsequent registrations for *Sempervivum* (and *Jovibara* and

Rosularia) were published by Mitchell (1982, 1983, 1985). Trehane (1989) also lists the cultivars of the houseleeks.

Serruria See *Proteaceae*.

Sinningia See *Gesneriaceae*.

Skimmia The cultivars of *Skimmia* are discussed by Bean (1970–1988), Brown (1980), Laar (1984), and Krüssmann (1984–1986).

Smithiantha See *Gesneriaceae*.

Solenophora See *Gesneriaceae*.

Sophora The cultivars of the Japanese pagoda tree, *S. japonica* L., are listed by Bean (1970–1988), Krüssmann (1984–1986), and Schalk (1985).

Sorbus The cultivars of the mountain ashes are discussed by Anonymous (1965), Bean (1970–1988), Krüssmann (1984–1986), Müssell (1971), Wright (1981), and Wyman (1969b). Hensen (1970) discusses the history and performance of cultivars of the *S. latifolia* (Lam.) Pers. complex. Huttleston (1990) records an additional registration.

Spathiphyllum The cultivars of the spathiphyllums are briefly listed by Chase et al. (1984).

Spiraea The cultivars of the spireas are listed by Bean (1970–1988), Grootendorst (1977), and Krüssmann (1984–1986).

Spyridium See Australian & South African plants.

Streptocarpus See *Gesneriaceae*.

Styrax The cultivars of the snowbells are listed by Fontaine (1970b) and reviewed by Raulston (1992).

Syringa The cultivars of lilacs are thoroughly discussed in Fiala (1988). This should be used in conjunction with Bean (1970–1988), Belorusets (1990), Bilov, Shtanko, & Mikhailov (1974), Gromov (1963), Harding (1933), Kalva (1980, 1988), Kelsey & Dayton (1942), Krüssmann (1984–1986), Luneva, Mikhailov, & Sudakova (1989), McKelvey (1928), Meyer (1952), Rogers (1976), Rubtsov, Zhogoleva, & Lyapunova (1961), Starcs (1928), Vrugtman (1990c, 1991), and Wister (1927, 1942, 1943, 1953a). The latest inventory of Russian cultivars has been translated into English by the International Lilac Society (Rubtsov et al., 1982). Recently the registrations have been published in *HortScience* (Vrugtman, 1988, 1989a, 1989b, 1990a, 1994b), while Vrugtman (1988) and Wister (1963) summarize previous registrations of lilacs.

Tagetes The African, French, and signet marigold cultivars are assembled in checklists by Maatsch & Nolting (1970) and Nolting & Zimmer (1975c, 1981, 1987).

Taxus See conifers.

Telopea See Australian & South African plants and *Proteaceae*.

Tetratheca See Australian & South African plants.

Thuja See conifers.

Thymus Flannery (1982) records the cultivars of thyme in her thorough Ph.D. thesis.

Tilia Bean (1970–1988), Grootendorst (1970), Krüssmann (1984–1986), and Santamour & McArdle (1985b) discuss the cultivars of the lindens. The cultivars of the lindens are also listed by Muir (1984, 1988) and Wyman (1962e) but with few introduction dates and little background. Huttleston (1989) records additional registrations.

Trees, shrubs, and woody vines (broad–leaved) The best general references on the introduction dates and descriptions of many broad–leaved trees, shrubs, and vines have been Bean (1970–1988) and Krüssmann (1984–1986). Rehder (1940, 1949) also lists many forma epithets; because these are published before the first International Code of Nomenclature for Cultivated Plants in 1952, the forma epithets are now considered cultivar names. Other woody species are listed by

Buckley (1980), *Commissie voor de samenstelling van de Rassenlijst voor Bosbouwgewassen* (1990), Darthuizer Boomkwekerijen B. V. (1987), Dirr (1990), Hillier (1982, 1991), Laar (1989), and Wyman (1963a, 1963b, 1966, 1967, 1969a). The mimeographed *Swarthmore Plant Notes* (Wister, 1954) are a treasure trove of information on cultivars of woody plants but, unfortunately, are not widely distributed. Registrations of recent woody genera have been published in *HortScience* (Huttleston, 1986, 1988, 1989, 1990; Spongberg, 1988, 1989, 1990), while Huttleston (1986) summarizes previous registrations. Cultivars of street trees are summarized by Gerhold et al. (1989) and Wandell (1989). Some trees and shrubs are also discussed in Hogan (1988) and the journal *Dendroflora* (see the cumulative indices in numbers 20 and 25).

Trichantha   See *Gesneriaceae*.

Trollius   The cultivars of the globe flowers are listed by Clausen (1973b) and Hensen (1959). Jelitto & Schacht (1990) and Trehane (1989) also list cultivars.

Tropaeolum   A list of cultivars of the common nasturtium is Kelsey & Dayton (1942) but without dates of introduction.

Tsuga   See conifers.

Tulipa   The cultivars of tulips are first listed by the Royal Horticultural Society (1908c) with significant revisions in 1917, 1929, 1930a, and 1939. Later, in 1948 and 1952, the Royal Horticultural Society published revisions in conjunction with the General Dutch Bulbgrowers Society. Later the Royal General Bulbgrowers' Society (1958, 1960, 1965, 1969) and the Royal General Bulbgrowers' Association (1971, 1976, 1981) published their own lists. The 1958–1965 editions are comprehensive, but the 1969–1981 editions only provided brief descriptions and dates for many tulips, and "historical cultivars" are appended at the rear of the list with no dates or descriptions. Kelsey & Dayton (1924), Kudryavtseva (1987), and Trehane (1989) are also recommended.

Ulmus   The cultivars of the elms are listed by Bean (1970–1988), Fontaine (1968), Green (1964), Krüssmann (1984–1986), and Touw (1963). Spongberg (1988, 1991) records additional registrations.

Variegated plants   The only work on cultivars of variegated plants is Yokoi & Hirose (1978). While the text is in Japanese, plant names are in English.

Veronica   The history and performance of cultivars of *Veronica* are discussed by Clausen (1971). Jelitto & Schacht (1990) and Trehane (1989) also list the cultivars of *Veronica*.

Viburnum   The cultivars of the viburnums are listed by Bean (1970–1988), Egolf (1968), and Krüssmann (1984–1986). Grootendorst (1979b) discusses the cultivars of *Viburnum plicatum*.

Vinca   Barnes (1984) and Hensen (1980) discuss the history and performance of the cultivars of *Vinca*. Jelitto & Schacht (1990) and Trehane (1989) also list cultivars.

Viola   Jelitto & Schacht (1990) and Trehane (1989) list the cultivars of *Viola*. The history and performance of the cultivars of *Viola cornuta* L., the viola, are discussed by Clausen (1969) and the Royal Horticultural Society (1912, 1913a). The cultivars of violets are treated comprehensively by Coombs (1981). Cultivars of pansies, violas, and violettas are listed by Fuller (1990) without dates or introducers.

Weigela   The cultivars of the weigelas are listed by Grootendorst (1968c), Howard (1965), and Schneider (1930). Bean (1970–1988) and Krüssmann (1958a, 1984–1986) also discuss cultivars. Spongberg (1988) records an additional registration.

Wisteria   Bowden (1976) and Browse (1984) review the available cultivars of the wisterias but without introduction dates or background. Sprenger (1911) lists the cultivars of *W. sinensis* (Sims) Sweet. Bean (1970–1988), Grootendorst (1968d), and Krüssmann (1984–1986) also discuss cultivars. Huttleston (1988) records an additional registration.

Zelkova The cultivars of the zelkovas are listed by Bean (1970–1988), Dirr (1990), Fontaine (1970c), and Krüssmann (1984–1986).

Zephyranthes The cultivars of the rain lilies are listed by Anonymous (1958e).

Zinnia The modern cultivars of *Zinnia* are listed, with some history, by Sharma & Metcalf (1968).

LITERATURE CITED

Aden, P., ed. 1990. *The Hosta book.* 2d ed. Portland, Oregon: Timber Press.

Agnew, N. H., and H. J. Lang. 1992. Outstanding New Guinea impatiens. *Greenhouse Manager* 10(10): 66–70.

Ahlburg, M. S. 1993. *Hellebores: christmas rose, lenten rose.* Trans. M. S. Ahlburg and J. Hewitt. London: B. T. Batsford.

Ahrendt, L. W. A. 1942. An analysis of the Wisley hybrid *Berberis. J. Roy. Hort. Soc.* 67: 129–135.

———. 1949. The *Berberis stenophylla* hybrids. *J. Roy. Hort. Soc.* 74: 36–40.

———. 1961. *Berberis* and *Mahonia*: A taxonomic revision. *J. Linn. Soc., Bot.* 57: 1–410.

Allan, P. 1988. Naming of cultivated plants. *Protea News* 7: 18–22.

Allen, I. 1983. Adventures with asters. *Hardy Pl. Soc. Bull.* 6(3): 75–79.

Allen, R. C. 1965. *Modern roses 6.* Harrisburg, Pennsylvania: J. Horace MacFarland.

Althofer, G. W. 1978. *Cradle of incense: The story of Australian Prostanthera.* Soc. Growing Austral. Pl.

American Association of Nurserymen. 1957. *Plant patents: common introductory names 1 through 1542.* Washington, DC: Am. Assoc. Nurserymen.

———. 1958. *1957 Supplement to plant patents: common introductory names 1543 through 1671.* Washington, DC: Am. Assoc. Nurserymen.

———. 1959. *1958 Supplement to plant patents: common introductory names 1672 through 1791.* Washington, DC: Am. Assoc. Nurserymen.

———. 1960. *1959 Supplement to plant patents: common introductory names 1792 through 1892.* Washington, DC: Am. Assoc. Nurserymen.

———. 1961. *1960 Supplement to plant patents: common introductory names 1893 through 2007.* Washington, DC: Am. Assoc. Nurserymen.

———. 1962. *1961 Supplement to plant patents: common introductory names 2009 through 2116.* Washington, DC: Am. Assoc. Nurserymen.

———. 1963. *Plant patents with common names 1 through 2207. 1931–1962.* Washington, DC: Am. Assoc. Nurserymen.

———. 1967. *Supplement to plant patents with common names 2699 through 2783.* Washington, DC: Am. Assoc. Nurserymen.

———. 1969. *Plant patents with common names 2208 through 2855. 1963–1968.* Washington, DC: Am. Assoc. Nurserymen.

———. 1974. *Plant patents with common names 2856 through 3412. 1969–1973.* Washington, DC: Am. Assoc. Nurserymen.

———. 1981. *Plant patents with common names 3413–4359. 1974–1978.* Washington, DC: Am. Assoc. Nurserymen.

———. c. 1987. *How to use, select, and register cultivar names.* Washington, DC: Am. Assoc. Nurserymen.

———. 1990. *Plant patent directory.* Washington, DC: Am. Assoc. Nurserymen.

American Begonia Society. 1957. *Buxton check list of begonias*. Los Angeles: Am. Begonia Soc.

———. 1958. *Buxton check list of begonias. Suppl. no. 1*. Los Angeles: Am. Begonia Soc.

———. 1962. *Buxton check list of begonias. Suppl. no. 2*. Los Angeles: Am. Begonia Soc.

———. 1967. *Buxton check list of begonias. Suppl. no. 3*. Los Angeles: Am. Begonia Soc.

———. 1985. *Catalog of registered cultivars of the genus Begonia. Vol. 1. Registration Numbers 1–100*. Encinitas, California: Am. Begonia Soc.

American Daffodil Society. 1977. *Daffodils to show and grow*. New Canaan, Connecticut: Am. Daffodil Soc.

———. 1985. *Daffodils to show and grow and abridged classified list of daffodil names*. 3rd ed. Hernando, Mississipi: Am. Daffodil Soc.

———. 1989. *Daffodils to show and grow and abridged classified list of daffodil names*. 4th ed. Hernando, Mississipi: Am. Daffodil Soc.

American Dahlia Society. 1989. *1989 Classification and handbook of dahlias*. Am. Dahlia Soc.

American Gesneria Society. 1957. Names and descriptions of plants currently in the trade. *Achimenes. Episcia. ×Gloxinera. Smithiantha*. Gesneriad Reg. [publ. with *Gesneriad J.* 4(4)].

American Gladiolus Society. 1931. *Descriptive gladiolus nomenclature*. Am. Gladiolus Soc.

American Gloxinia Society. 1962. Names and descriptions of plants commercially available. *Kohleria. Smithiantha*. Gesneriad Reg. [publ. with *Gloxinian* 12(5)].

American Hemerocallis Society. 1957. *Hemerocallis check list 1893 to July 1, 1957*. Am. Hemerocallis Soc.

———. c. 1973. *Hemerocallis check list July 1, 1957 to July 1, 1973*. Am. Hemerocallis Soc.

———. c. 1984. *Hemerocallis check list July 1, 1973 to December 31, 1983*. Am. Hemerocallis Soc.

American Hibiscus Society. 1984. *Hibiscus catalogue*. Am. Hibiscus Soc.

———. 1987. *Nomenclature of the American Hibiscus Society*. Am. Hibiscus Soc.

Anderson, A. J. 1971. *Osmunda regalis* and its forms. *Gard. Chron.* 169(11): 32.

Andrews, J. 1984. *Peppers: the domesticated capsicums*. Austin: Univ. Texas Press.

Andrews, S. 1983. Notes on some *Ilex ×altaclerensis* clones. *Plantsman* 5: 64–81.

———. 1984. More notes on clones of *Ilex ×altaclerensis*. *Plantsman* 6: 157–166.

———. 1986. The clones of *Ilex ×altaclerensis*. *Acta Hort.* 182: 377–380.

Anonymous. [Not dated] *Willow varieties for Alaska*. Palmer, Alaska: Plant Mater. Center.

———. 1958a. *Oleander. Bull. Lucknow Natl. Bot. Gard.* 10: 1–8.

———. 1958b. *Plumeria. Bull. Lucknow Natl. Bot. Gard.* 14: 1–11.

———. 1958c. *Ixora. Bull. Lucknow Natl. Bot. Gard.* 15: 1–14.

———. 1958d. *Lantana. Bull. Lucknow Natl. Bot. Gard.* 19: 1–12.

———. 1958e. *Cooperanthes* and *Zephyranthes. Bull. Lucknow Natl. Bot. Gard.* 20: 1–34.

———. 1958f. *Amaryllis. Bull. Lucknow Natl. Bot. Gard.* 21: 1–27.

———. 1959a. *Codiaeum* (Croton). *Bull. Lucknow Natl. Bot. Gard.* 31: 1–24.

———. 1959b. *Cosmos* (Cosmea). *Bull. Lucknow Natl. Bot. Gard.* 34: 1–16.

———. 1959c. *Bougainvillea. Bull. Lucknow Natl. Bot. Gard.* 41: 1–35.

———. 1959d. Cordylines & dracaenas. *Bull. Lucknow Natl. Bot. Gard.* 42: 1–18.

——. 1960. *Nymphaea, Nelumbo, Euryale* and *Victoria*. *Bull. Lucknow Natl. Bot. Gard.* 50: 1–20.

——. 1965. *Sorbus. Dendroflora* 2: 28–44.

Aril Society. 1976. *The Aril Society international official checklist*. Aril Soc.

——. 1978. *The Aril Society international official checklist. add. no. 1*. Aril Soc.

——. 1980. *The Aril Society international official checklist. add. no. 2*. Aril Soc.

——. 1982. *The Aril Society international official checklist. add. no. 3*. Aril Soc.

Arnold, P. 1963a. Names and descriptions of plants commercially available in the genus *Episcia*. Gesneriad Reg. [publ. with *Gloxinian* 13(1)].

——. 1963b. Names and descriptions of plants commercially available in the genus *Columnea*. Gesneriad Reg. [publ. with *Gloxinian* 13(6)].

——. 1966. Names and descriptions of plants commercially available in the genus *Columnea*. Gesneriad Reg. [publ. with *Gloxinian* 16(3)].

——. 1968. Names and descriptions of plants commercially cultivated in the genus *Episcia*. Gesneriad Reg. [publ. with *Gloxinian* 18(1)].

——. 1969. Names and descriptions of plants commercially available in the genus *Achimenes* and registered cultivar names. Gesneriad Reg. [publ. with *Gloxinian* 19(5)].

——. 1972. Names and descriptions of cultivated plants in the genus *Streptocarpus*. Gesneriad Reg. [publ. with *Gloxinian* 22(5)].

——. 1975. Check list of names with descriptions of cultivated plants in the genus *Sinningia*. Gesneriad Reg. [publ. with *Gloxinian* 25(2)].

——. 1977. Check list of names with descriptions of cultivated plants in the genus *Episcia*. Gesneriad Reg. [publ. with *Gloxinian* 27(2)].

——. 1978. Check list of names with description of cultivated plants in the genus *Nematanthus*. Gesneriad Reg. [publ. with *Gloxinian* 28(5)].

——. 1979. Check list of names with descriptions of cultivated plants in the genus *Streptocarpus*. Gesneriad Reg. [publ. with *Gloxinian* 29(6)].

Ashburner, K. 1986. *Alnus*—A survey. *Plantsman* 8: 170–188.

——. 1980. *Betula*—A survey. *Plantsman* 2: 31–53.

Austin, D. 1988. *The heritage of the rose*. Woodbridge, England: Antique Collectors' Club.

Australian Cultivar Registration Authority. 1988. *Registered cultivars*. Austral. Cultivar Registration Authority.

Australian Geranium Society. 1978. *A check list and register of Pelargonium cultivar names. Pt. 1, A–B*. Sydney: Austral. Geranium Soc.

——. 1985. *A check list and register of Pelargonium cultivar names. Pt. 2, C–F*. Sydney: Austral. Geranium Soc.

Bachtell, K. R., and E. R. Hasselkus. 1982. Selecting varieties of shrub *Potentilla* for the Midwest. *Am. Nurseryman* 155(3): 83–89.

Bagust, H. 1988. *Miniature and dwarf geraniums (pelargoniums)*. Portland, Oregon: Timber Press.

Bailey, S. 1990. *Carnations*. London: Blandford.

Barnes, P. 1984. Trial of *Vinca* cultivars. *The Garden* (London) 109: 426–429.

Barret, P. W. 1959. *Aster ericoides*. *Hardy Pl. Soc. Bull.* 2(1): 17–18.

Bartlett, M. 1975. *Gentians*. Poole, England: Blandford Press.

Batcheller, F. N. 1985. Check list of names with descriptions of cultivated plants in the genus *Kohleria*. Ed. J. D. Dates. Gesneriad Reg. [publ. with *Gloxinian* 35(5)].

Batdorf, L. R. 1987. International registration list of cultivated *Buxus* L. *Boxwood Bull.* 26: 76–81.

——. 1988. Checklist of *Buxus* L. *Boxwood Bull.* 28: 43–49.

Beadle, D. A. 1991. *A preliminary listing of all known cultivar and grex names for the Bromeliaceae.* Corpus Christi, Texas: Bromeliad Soc.
Beal, A. C. 1920. *The peony: a flower for the farmer.* Cornell Reading Course for the Farm, Lesson 154. Ithaca: New York State College of Agriculture, Cornell University.
Beales, P. 1985. *Classic roses.* New York: Holt, Rinehart and Winston.
———. 1988. *Twentieth-century roses.* New York: Harper & Row.
Bean, W. K. 1970–1988. *Trees and shrubs hardy in the British Isles.* 4 vols. + suppl. London: John Murray.
Beckett, K. A. 1983. *Bergenia. The Garden* (London) 108: 480–484.
Beers, L., and J. Howie. 1985. *Growing Hibiscus.* Kenthurst, Australia: Kangaroo Press.
———. 1990. *Growing Hibiscus.* 2d ed. Kenthurst, Australia: Kangaroo Press.
Belorusets, E. Sh. 1990. *Siren'* (in Russian). Kiev: Uroshai.
Berry, F., and W. J. Kress. 1991. *Heliconia: an identification guide.* Washington, DC: Smithsonian Inst. Press.
Bilov, V. E., I. I. Shtanko, and N. L. Mikhailov. 1974. *Siren'* (in Russian). Moscow: Hayka.
Birch, C. H. 1940. Parentages of named varieties. *Gladiolus* 1940: 33–40.
Bishop, F. 1949. *The delphinium: a flower monograph.* London: Collins.
Blasdale, W. C. 1948. *The cultivated species of Primula.* Berkeley: Univ. California Press.
Boitard, P. 1836. *Manuel complet de l'amateur de roses.* Paris: Libr. Encycl. Roret.
Boland, M. 1983. *African violet master variety list 1976–1983. Registered varieties 1948–1983.* Knoxville, Tennessee: African Violet Soc. Am.
———. 1984. The 1984 master list of African violets. *African Violet Mag.* 37(4): 43–55.
———. 1985. The 1985 master list of African violets. *African Violet Mag.* 38(4): Suppl.
———. 1986. The 1986 master list of African violets. *African Violet Mag.* 39(5): 31–42.
———. 1987. The 1987 master list of African violets. *African Violet Mag.* 40(6): 27–45.
———. 1988. The 1988 master list of African violets. *African Violet Mag.* 41(6): Suppl.
Bom, P. L. M. van der. 1982. Amerikaanse selecties van straat, laan en parkbomen. *Dendroflora* 19: 44–82.
Bond, J. 1982. *Pieris*—A survey. *Plantsman* 4: 65–75.
Boullemier, L. B. 1975. *A check list of species, hybrids and cultivars of the genus Fuchsia with 1976 Add. no. 1, 1978 Add. no. 2.* Poole, England: Blandford Press.
———. 1980. *Add. no. 3 to a check list of species, hybrids and cultivars of the genus Fuchsia.* Poole, England: Blandford Press.
———. 1982. *Add. no. 4 to a check list of species, hybrids and cultivars of the genus Fuchsia.* Poole, England: Blandford Press.
———. 1985. *The checklist of species, hybrids and cultivars of the genus Fuchsia.* Poole, England: Blandford Press.
Bourne, S. E. 1903. *The book of the daffodil.* London: John Lane.
Bowden, W. M. 1957. Cytotaxonomy of *Potentilla fruticosa*, allied species and cultivars. *J. Arnold Arbor.* 38: 381–388.
———. 1976. A survey of wisterias in southern Ontario gardens. *Roy. Bot. Gard. Techn. Bull.* No. 8.

——. 1983. *Descriptions of eighteen tetraploid Lobelia cultivars.* Simcoe, Ontario: Wray M. Bowden.

Bowers, C. G. 1960. *Rhododendrons and azaleas: their development, cultivation and development.* New York: Macmillan.

Bowles, E. A. 1934. *A handbook of Narcissus.* London: Martin Hopkinson.

——. 1956. Garden varieties of *Galanthus.* Pp. 65–70 in F. C. Stern, *Snowdrops and snowflakes.* London: Roy. Hort. Soc.

Brearley, C. 1987. The shrubby potentillas. *Plantsman* 9: 90–109.

Brickell, C. D. 1979. The hybrids between *Mahonia japonica* and *M. lomariifolia. Plantsman* 1: 12–20.

——. 1980. *Rhododendron* cultivars, their nomenclature and international registration. Pp. 63–73 in J. L. Luteyn, ed., *Contributions toward a classification of Rhododendron.* New York: New York Bot. Gard.

Brickell, C. D., and B. Mathew. 1976. *Daphne: the genus in the wild and in cultivation.* Woking, England: Alpine Garden Soc.

British Pelargonium and Geranium Society. 1970. *BPGS check list: Erodium.* British Pelargonium and Geranium Soc.

Brits, G. J. 1988a. Cultivar Registration: International *Protea* cultivar registration. *Protea News* 7: 3–5.

——. 1988b. Sample list of validly published *Protea* cultivar names. *Protea News* 7: 6–10.

——. 1988c. International naming and registration of *Protea* cultivars. Ed. P. Mathews. *Protea News* 7: 11–17.

Broekhuizen, J. T. M. van. 1977. Het geslacht *Populus*; taxonomie en overzicht van de belangrijkste soorten en hybriden. *Dendroflora* 13/14: 40–49.

Broekhuizen, J. T. M. van, and F. Schneider. 1969. *Salix alba* als laanboom. *Dendroflora* 6: 67–74.

Bromeliad Society. 1989. *Final tentative list of bromeliad cultivars.* Bromeliad Soc.

Brookins, L. 1991a. *The 1991 cumulative check list of reblooming Iris.* Menomonee Falls, Wisconsin: Am. Iris Soc., Reblooming Iris Soc.

——. 1991b. *The 1991 cumulative check list of Siberian Iris.* Menomonee Falls, Wisconsin: Am. Iris Soc., Soc. Siberian Irises.

——, ed. 1992. *The 1992 cumulative check list of Japanese irises.* Japanese Iris. Soc.

Brown, A. G. 1973. Hybrid *Streptocarpus. J. Roy. Hort. Soc.* 98: 201–205.

Brown, B. F. 1960. *Florida's beautiful crotons.* Am. Codiaeum Soc.

Brown, P. D. 1980. The genus *Skimmia* as found in cultivation. *Plantsman* 1: 224–249.

Browse, P. M. 1984. Some notes on members of the genus *Wisteria* and their propagation. *Plantsman* 6: 109–122.

Buckley, A. R. 1980. *Trees and shrubs of the Dominion Arboretum.* Res. Branch, Agric. Canad. Dept. Publ. 1697.

Bulgin, L. W. 1986. *Rhododendron hybrids: a compendium by parent.* Sherwood, Oregon: Ellanhurst Garden.

Byrne, N. J., 1986. Industrial patents for plant cultivars and breeding methods. Pp. 385–391 in B. T. Styles, ed., *Infraspecific classification of wild and cultivated plants.* Oxford: Clarendon Press.

Caillet, M., and J. K. Metzweiller. 1988. *The Louisiana Iris: the history and culture of five native American species and their hybrids.* Soc. Louisiana Irises.

Cairns, E., ed. 1993. *Modern roses 10.* Shreveport, Louisiana: Am. Rose Soc.

Carruthers, L., and R. Ginns. 1973. *Echeverias.* New York: Arco Publ.

Chadbund, G. 1972. *Flowering cherries.* London: Collins.

Chadwick, L. C., and R. A. Keen. 1976. *A study of the genus Taxus*. Ohio Agric. Res. Developm. Center Res. Bull. 1086.

Chahinian, B. J. 1986. *The Sansevieria trifasciata varieties*. Reseda, California: Trans Terra Publ.

Chalk, D. 1988. *Hebes and parahebes*. Portland, Oregon: Timber Press.

Chapple, F. J. 1951. *The heather garden*. London: W. H. & L. Collingridge.

Chase, A. R., R. T. Poole, L. S. Osborne, and R. J. Henny. 1984. *Spathiphyllum*. *Foliage Dig.* 7(7): 6–8.

Cheek, R. 1979. New Zealand flax. *The Garden* (London) 104: 101–106.

———. 1993. La belle marguerite. *The Garden* (London) 118: 350–355.

Chin, H. F. 1986. *The hibiscus: queen of tropical flowers*. Kuala Lumpur: Trop. Press.

Chinn, J. F., and R. A. Criley. 1982. *Plumeria cultivars in Hawaii*. Hawaii Agr. Exp. Sta. Res. Bull. 158.

Choudhary, B., and B. Singh. 1981. *The international Bougainvillea check list*. New Delhi: Indian Agric. Res. Inst.

Clausen, G. 1968. Sortsafprøvning af *Helianthemum* 1963–66. *Tiddsskr. Planteavl* 71: 512–517.

———. 1969. Sortsforsøg med vegetativt og frøformerede *Viola cornuta* 1965–68. *Tiddsskr. Planteavl* 73: 434–450.

———. 1970. Sortsforsøg med *Heuchera* 1964–69. *Tiddsskr. Planteavl* 74: 111–116.

———. 1971. Arts og sortsforsøg med *Veronica* 1966–69. *Tiddsskr. Planteavl* 75: 183–190.

———. 1972a. Sortsforsøg med japanske anemoner 1965–70. *Tiddsskr. Planteavl* 76: 13–21.

———. 1972b. Sortsforsøg med *Erigeron* 1969–71. *Tiddsskr. Planteavl* 76: 511–518.

———. 1973a. Forsøg med sorter af *Aster amellus* L., *Aster dumosus* L., *Aster ericoides* L., *Aster vimineus* Lam., *Aster novae-angliae* L. og *Aster novi-belgii* L. *Tiddsskr. Planteavl* 77: 19–36.

———. 1973b. Sortsforsøg med *Trollius* 1968–1972. *Tiddsskr. Planteavl* 77: 429–438.

———. 1973c. Sortsforsøg med *Aubrieta* Adans. 1970–1972. *Tiddsskr. Planteavl* 77: 587–607.

———. 1974a. Sortsforsøg med *Heliopsis* 1969–71. *Tiddsskr. Planteavl* 78: 429–434.

———. 1974b. Arts–og sortsforsøg med *Geranium* 1971–73. *Tiddsskr. Planteavl* 73: 441–468.

———. 1974c. Sortsforsøg med *Helianthus* 1969–72. *Tiddsskr. Planteavl* 78: 435–440.

———. 1975. Sortsforsøg med *Geum* 1971–73. *Tiddsskr. Planteavl* 79: 1–9.

———. 1976. Sortsforsøg med *Campanula* 1970–74. *Tiddsskr. Planteavl* 80: 443–461.

———. 1978. Arts og sortsforsøg med *Sedum* 1973–76. *Tiddsskr. Planteavl* 82: 145–164.

Clifford, D. 1970. *Pelargoniums including the popular "geranium"*. London: Blandford Press.

Clifton, R. T. F. 1992. The *Geraniaceae* group. Geranium family species check list. 4th ed. pt. 2. *Geranium*. Dover, England: *Geraniaceae* Group.

Coit, J. E. 1907. *A peony check-list*. Ithaca: New York State Coll. Agric., Cornell Univ.

Collings, W. J. 1986. *Supplement to named lily hybrids and their origins*. North Am. Lily Soc.

Commissie voor de samenstelling van de Rassenlijst voor Bosbouwgewassen. 1990. *5ᵉ Rassenlijst van bomen: Fifth list of approved clones and provenances of trees*. Maastricht: Leiter–Nypels V.B.

Conard, H. S. 1905. *The waterlilies: a monograph of the genus Nymphaea*. Publ. Carnegie Inst. Wash. No. 4.

Coombes, A. J. 1988. *Phygelius* in the wild and in cultivation. *Plantsman* 9: 233–246.
Coombs, R. E. 1981. *Violets*. Kent, England: Croom Helm.
Cooper, L. 1984. The international registration of *Delphinium* cultivar names. *Delphinium Soc. Yearb.* 1984: 116–118.
——. 1986. The international *Delphinium* register supplement 1984–1985. *Delphinium Soc. Yearb.* 1986: 63–64.
——. 1987. The international *Delphinium* register supplement 1985–1986. *Delphinium Soc. Yearb.* 1987: 104–105.
——. 1989. The international *Delphinium* register supplement 1987–88. *Delphinium Soc. Yearb.* 1989: 101.
——. 1990a. *A plantsman's guide to delphiniums*. London: Ward Lock.
——. 1990b. The international *Delphinium* register supplement 1988–89. *Delphinium Soc. Yearb.* 1990: 107.
——. 1991. The international *Delphinium* register supplement 1989–90. *Delphinium Soc. Yearb.* 1991: 86.
——. 1992. The international *Delphinium* register supplement 1990–91. *Delphinium Soc. Yearb.* 1992: 77.
——. 1993. The international *Delphinium* register supplement 1991–92. *Delphinium Soc. Yearb.* 1993: 70–71.
——. 1994. The international *Delphinium* register supplement 1992–93. *Delphinium Soc. Yearb.* 1994: 77–78.
Cope, E. A. 1986. *Native and cultivated conifers of northeastern North America.* Ithaca, New York: Cornell Univ. Press.
Cox, P. A. 1985. *The smaller rhododendrons*. Portland, Oregon: Timber Press.
Cox, P. A., and K. Cox. 1988. *Encyclopedia of rhododendron hybrids*. Portland, Oregon: Timber Press.
Cumming, R. W. 1964. *The chrysanthemum book*. Princeton, New Jersey: D. Van Nostrand.
Darke, R. 1990. *Idea garden ornamental grasses*. Kennett Square, Pennsylvania: Longwood Gardens.
Darlington, C. D., J. B. Hair, and R. Hurcombe. 1951. The history of the garden hyacinths. *Heredity* 5: 233–252.
Darrow, G. M., and F. G. Meyer, eds. 1968. Daylily handbook. *Am. Hort. Mag.* 47: 41–272.
Darthuizer Boomkwekerijen B. V. 1987. *Darthuizer Vademecum*. 3d rev. ed. Leersum, Netherlands: Darthuizer Boomkwekerijen B. V.
Dates, J. D. 1986. Check list of names with descriptions of intergeneric hybrids in the tribe Gloxinaieae (Gloxinieae). Registered Gesneriad List 1957–1986. Gesneriad Reg. [publ. with *Gloxinian* 36(6)].
——. 1987. Check list of names with descriptions of cultivated plants in the genera *Bucinellina, Columnea, Dalbergaria, Pentadenia*, and *Trichantha* also intergeneric hybrids. Gesneriad Reg. [publ. with *Gloxinian* 37(6)].
——. 1988. Check list of names with descriptions of cultivated plants in the genus *Sinningia*. Gesneriad Reg. [publ. with *Gloxinian* 39(1)].
——. 1990. Check list of names with descriptions of cultivated plants in the genus *Aeschyanthus*. Gesneriad Reg. [publ. with *Gloxinian* 40(6)].
——. 1993. Check list of names with descriptions of cultivated plants in the genera *Episcia* and *Alsobia*. Gesneriad Reg.
Davies, D. 1992. *Alliums: the ornamental onions*. Portland, Oregon: Timber Press.
Debor, H. W. 1978. Bibliographie des internationalen haselnuss–schrifttums. *Bibliogr. Reihe Techn. Univ. Berlin*. Vol. 10. Berlin: Universitatsbibliothek Techn. Univ.

den Boer, A. F. 1959. *Ornamental crab apples*. Washington, DC: Am. Assoc. Nurserymen.

den Ouden, P., and B. K. Boom. 1978. *Manual of cultivated conifers hardy in the cold– and warm–temperate zone*. The Hague: Martinus Nijhoff.

Desportes, N. H. F. 1828. *Rosetum Gallicum*. Paris: Pesche.

de Vos, F. 1958. Cultivated firethorns. *Proc. Pl. Prop. Soc.* 8: 32–38.

Dickerson, B. C. 1992. *The old rose advisor*. Portland, Oregon: Timber Press.

Dirr, M. A. 1988. To know them is to love them: Fruited, deciduous hollies can extend color, charm and profits. *Am. Nurseryman* 168(3): 22–28, 32–41.

———. 1990. *Manual of woody landscape plants: their identification, ornamental characteristics, culture, propagation and uses*. 4th ed. Champaign, Illinois: Stipes.

Dobson, B. R. 1987. *Combined rose list 1987*. Irvington, New York: B. R. Dobson.

———. 1988. *Combined rose list 1988*. Irvington, New York: B. R. Dobson.

———. 1989. *Combined rose list 1989*. Irvington, New York: B. R. Dobson.

———. 1990. *Combined rose list 1990*. Irvington, New York: B. R. Dobson.

———. 1991. *Combined rose list 1991*. Irvington, New York: B. R. Dobson.

Dobson, B. R., and P. Schneider. 1992. *Combined rose list 1992*. Rocky River, Ohio: Peter Schneider.

Dolatowski, J. 1986. Cultivars of mock orange cultivated in Poland. *Arbor. Kórnickie* 31: 39–62.

Donald, K. 1986. The Royal Horticultural Society: Its role as the International Registration Authority for Narcissus. *Acta Hort.* 182: 381–386.

Douglas, G., ed. 1949. *Alphabetical Iris check list*. Nashville: Am. Iris Soc.

Dress, W. J. 1962. Notes on cultivated Compositae. 7. *Ligularia*. *Baileya* 10: 62–87.

Dudley, T. R., 1966. Ornamental madworts *(Alyssum)* and the correct name of the goldentuft alyssum. *Arnoldia* 26: 33–45.

Dudley, T. R., and G. K. Eisenbeiss. 1971. Registration and documentation of cultivar names. *Boxwood Bull.* 11: 12–14.

———. 1992. *International checklist of cultivated Ilex. pt. 2. Ilex crenata Thunberg ex J. A. Murray*. Natl. Arbor. Contrib. No. 9.

Durrant, T. 1982. *The Camellia story*. Auckland: Heinemann.

Dwarf Iris Society. 1975. *Alphabetical checklist of miniature dwarf Iris: a cumulative index of small Iris*. Portland, Indiana: Dwarf Iris Soc.

———. 1988. *An alphabetical checklist of miniature dwarf Iris: an update of the cumulative index of small Iris published 1975*. Wichita, Kansas: Dwarf Iris Soc.

Ecke, P., ed. 1976. *The poinsettia manual*. Encinitas, California: Paul Ecke Poinsettias.

Edwards, C. 1989. *Delphiniums: the complete guide*. Marlborough, England: Crowood Press.

Edwards, J. 1987. A list of named cultivars in commerce. *Delphinium Soc. Yearb.* 1987: 106–109.

Eggenberger, R. M., and M. H. Eggenberger. 1988. *The handbook on Plumeria culture*. 2nd ed. Houston: Plumeria People.

Eggerss, M. L., and E. R. Hasselkus. 1992. The best of the winterberries. *Am. Nurseryman* 176(12): 115–125.

Egolf, D. R. 1968. *The cultivated viburnums*. Washington, DC: U.S. National Arboretum.

Egolf, D. R., and A. O. Andrick. 1978. *The Lagerstroemia handbook/checklist: a guide to crapemyrtle cultivars*. Am. Assoc. Bot. Gard.

Eichin, R., and E. Deiser. 1988. Impatiens–Neu–Guinea–Hybriden. *Gartenpraxis* 1988(4): 14–17.
Eisenbeiss, G. K., and T. R. Dudley. 1973. *International checklist of cultivated Ilex. pt. 1. Ilex opaca.* Natl. Arbor. Contr. No. 3.
——. 1983. *Ilex cultivar registration list, 1958–1983.* Holly Soc. Am. Bull. No. 20.
Erdman, R. P. 1949. *American Camellia catalog.* Savannah, GA: R. P. Erdman.
Erhardt, W. 1992. *Hemerocallis: day lilies.* Portland, Oregon: Timber Press.
Evans, R. L. 1983. *Handbook of cultivated sedums.* London: Sci. Rev.
Evison, R. 1985. Large–flowered *Clematis* cultivars. *The Garden* (London) 110: 207–213.
Ewart, R. 1982. *Fuchsia lexicon.* New York: Van Nostrand Reinhold.
——. 1987. *Fuchsia lexicon.* Rev. ed. London: Blandford.
Fagan, G. 1988. *Roses at the Cape of Good Hope.* Cape Town: Breestraat–Publikasies.
Fearnley–Whittingstall, J. 1992. *Ivies.* New York: Random House.
Fiala, J. L. 1988. Lilacs: *The genus Syringa.* Portland, Oregon: Timber Press.
Fisher, E. V. 1978. *Named lily hybrids and their origins.* Ed. J. Montgomery. North Am. Lily Soc.
——. 1979. *Hosta: the aristocratic plant for shady gardens.* Am. Hosta Soc.
Fisher, S. N., Lt Col. & Mrs, and Mr & Mrs Robert Ellis. 1975. *Gladiolus variety parentages.* 1975. Rev. ed. North Am. Gladiolus Council.
——. c. 1983. *Gladiolus variety parentages.* North Am. Gladiolus Council.
Flannery, H. B. 1982. A study of the taxa of *Thymus* L. (*Labiatae*) cultivated in the United States. Ph.D. Thesis. Ithaca, New York: Cornell Univ.
Fleming, R. 1979. Hybrid *Nepenthes. Carnivorous Pl. Newslett.* 8: 10–12.
Fletcher, H. R. 1958. *The international Rhododendron register.* London: Roy. Hort. Soc.
Fogg, J. M., and J. C. McDaniel. 1975. *Check list of the cultivated magnolias.* Mt Vernon, Virginia: Am. Hort. Soc.
Fontaine, F. J. 1968. *Ulmus;* Keuringsrapport van de Regelingskommissie Sierbomen N.A.K.B. *Dendroflora* 5: 36–55.
——. 1970a. Het geslacht *Betula. Meded. Bot. Tuinen Belmonte Arbor. Wageningen* 13: 99–180.
——. 1970b. *Styrax, Pterostyras, Halesia. Dendroflora* 7: 63–68.
——. 1970c. *Zelkova. Dendroflora* 7: 82–84.
Foreman, J. J. 1985. *Spuria irises: introduction and varietal listing.* James J. Foreman.
Frank, H. 1975. *African violet miniature and semiminiature variety list. 4th ed. Classification supplement—1976, 1977.* African Violet Soc. Am.
Fretwell, B. 1989. *Clematis.* Deer Park, Wisconsin: Capability's Books.
Fritchie, C. 1982. *Louisiana Iris cultivars.* Soc. Louisiana Irises.
Fuller, R. 1990. *Pansies, violas & violettas: the complete guide.* Marlborough, England: Crowood Press.
Galle, F. C. 1985. *Azaleas.* Portland, Oregon: Timber Press.
Gardiner, J. M. 1989. *Magnolias.* Chester, Connecticut: Globe Pequot Press.
Gault, S. M., and P. M. Synge. 1971. *The dictionary of roses in color.* New York: Grosset and Dunlap.
Gelderen, D. M. van. 1971. *Ilex. Dendroflora* 8: 19–35.
——. 1975. *Picea* en *Abies;* Treur– en Kruipvormen. *Dendroflora* 11/12: 38–45.
——. 1979. *Pieris. Dendroflora* 15/16: 36–44.
——. 1982. *Pinus. Dendroflora* 19: 3–28.
——. 1984. *Juniperus;* breed en plat groeiende cultivars. *Dendroflora* 21: 2–38.
——. 1988. *Ilex aquifolium* en *Ilex* ×*altaclerensis. Dendroflora* 25: 7–34.

Gelderen, D. M. van, and J. R. P. van Hoey Smith. 1986. *Conifers.* Portland, Oregon: Timber Press.
——. 1992. *Rhododendron portraits.* Portland, Oregon: Timber Press.
Genders, R. 1961. *Miniature chrysanthemums and koreans.* New York: St. Martin's Press.
——. 1962. *Primroses.* New York: St. Martin's Press.
——. 1963a. *Delphiniums.* London: John Gifford.
——. 1963b. *The polyanthus.* London: Faber & Faber.
Gerbing, G. G. 1945. *Camellias. Series 2.* Loose leaf binder. G. G. Gerbing.
Gerhold, H. D., W. N. Wandell, N. L. Lacasse, and R. D. Schein, eds. 1989. *Street tree factsheets.* University Park: Penn. State Univ.
Gibbs, V. 1929. Robinias at Aldenham and Kew. *J. Roy. Hort. Soc.* 54: 145–158.
Gillis, W. T. 1976. Bougainvilleas of cultivation (*Nyctaginaceae*). *Baileya* 20: 34–41.
Gilmour, J. S. L., J. Greatwood, and P. F. Hunt. 1976. *Handbook on orchid nomenclature and registration.* Cambridge, Massachussets: Int. Orchid Commiss., Am. Orchid Soc.
Gioia, V. G. 1986. Revised rose name registration system. *Acta Hort.* 182: 265–271.
Goodwin, F. H. 1986. The control of plant variety rights. Pp. 375–383 in B. T. Styles, ed., *Infraspecific classification of wild and cultivated plants.* Oxford: Clarendon Press.
Gosling, S. G. 1964. *British national register of names of chrysanthemums.* 3d ed. London: National Chrysanthemum Soc.
——. 1973a. *The pocket encyclopedia of chrysanthemums.* New York: Arco Publ.
——. 1973b. *British national register of names of chrysanthemums.* Suppl. ed. 1964–1972. London: National Chrysanthemum Soc.
——. 1980. *British national register of names of chrysanthemums.* Amalgamated ed. 1964–1979. London: National Chrysanthemum Soc.
Goudey, C. J. 1985. *Maidenhair ferns in cultivation.* Portland, Oregon: Lothian Publ.
Graf, A. B. 1986a. *Exotica international.* Series 4. 12th ed. East Rutherford, New Jersey: Roehrs.
——. 1986b. *Tropica.* 3d ed. East Rutherford, New Jersey: Roehrs.
Gravereaux, J. 1902. *Les roses cultivées à l'Haÿ en 1902.* Paris.
Green, P. S. 1963. *Leucothoe fontanesiana. Arnoldia* 23: 93–99.
——. 1964. Registration of cultivar names in *Ulmus. Arnoldia* 24: 41–80.
——. 1965. Studies in the genus *Jasminum* III. The species in cultivation in North America. *Baileya* 13: 137–172.
Greer, H. E. 1982. *Greer's guidebook to available rhododendrons.* Eugene, Oregon: Offshoot Publ.
Grenfell, D. 1986. Some comments on Dr. Hensen's study of the taxonomy of cultivated hostas. *Plantsman* 7: 251–255.
——. 1990. *Hosta: the flowering foliage plant.* London: B. T. Batsford.
Grey-Wilson, C. 1988. *The genus Cyclamen.* Kew: Roy. Bot. Garden
——. 1993. *Poppies: a guide to the poppy family in the wild and in cultivation.* London: B. T. Batsford.
Griffiths, T. 1984. *The book of old roses.* London: Michael Joseph.
——. 1987. *The book of classic old roses.* London: Michael Joseph.
Gromov, A. 1963. *Siren'.* Moscow: Moskovskii Rabochii.
Grootendorst, H. J. 1954. *Rhodendrons en Azaleas.* Vereniging voor Boskoopse Culturen.
——. 1964a. *Malus*—Sierappels. *Dendroflora* 1: 2–15.
——. 1964b. *Prunus;* sierkersen; het sierkersen sortiment. *Dendroflora* 1: 16–31.
——. 1965. *Hamamelis. Dendroflora* 2: 11–17.

———. 1966a. Laag groeiende cotoneasters. *Dendroflora* 3: 20–27.
———. 1966b. *Fraxinus;* Keuringsrapport van de Regelingscommissie Sierbomen. N.A.K.B. *Dendroflora* 3: 28–36.
———. 1967a. *Aesculus;* Keuringsrapport van de Regelingscommissie Sierbomen N.A.K.B. *Dendroflora* 4: 2–9.
———. 1967b. Bladverliezende Azaleas. *Dendroflora* 4: 10–17, ill. opposite pp. 9, 16, 17, 56, 57.
———. 1967c. *Crataegus;* Keuringsrapport van de Regelingscommissie Sierbomen N.A.K.B. *Dendroflora* 4: 18–29.
———. 1968a. *Chaenomeles. Dendroflora* 5: 13–19.
———. 1968b. *Hibiscus syriacus. Dendroflora* 5: 23–28.
———. 1968c. *Juniperus communis;* opgaande vormen. *Dendroflora* 5: 29–34.
———. 1968c. *Weigela. Dendroflora* 5: 56–60, ill. opp. pp. 64, 65.
———. 1968d. *Wisteria. Dendroflora* 5: 61–68.
———. 1969a. *Acer;* Keuringsrapport van de Regelingskommissie Sierbomen N.A.K.B. *Dendroflora* 6: 3–18.
———. 1969b. Japanse Azaleas. *Dendroflora* 6: 37–43.
———. 1969c. *Rhododendron* tuinhybriden. *Dendroflora* 6: 57–66.
———. 1970. *Tilia;* Keuringsrapport van de Regelingskommissie Sierbomen N.A.K.B. *Dendroflora* 7: 69–81.
———. 1971a. *Robinia;* Keuringsrapport van de Keuringscommissie Sierbomen N.A.K.B. *Dendroflora* 8: 46–57.
———. 1971b. *Thuja occidentalis*–bolvormen. *Dendroflora* 8: 58–59.
———. 1972a. *Alnus;* Keuringsrapport van de Regelingskommissie Sierbomen N.A.KB. *Dendroflora* 9: 2–8.
———. 1972b. *Buddleja. Dendroflora* 9: 38–42.
———. 1973a. *Betula;* Keuringsrapport van de Keuringscommissie Sierbomen N.A.K.B. *Dendroflora* 10: 15–25.
———. 1973b. *Hydrangea. Dendroflora* 10: 26–40.
———. 1975. *Fagus. Dendroflora* 11/12: 3–17.
———. 1977. *Spiraea. Dendroflora* 13/14: 50–61.
———. 1979a. *Rhododendron williamsianum* hybriden. *Dendroflora* 15/16: 45–49.
———. 1979b. *Viburnum plicatum* en de cultuur-variëteiten. *Groen* 35: 361–362.
———. 1980a. *Hamamelis;* Keuringsrapport. *Dendroflora* 17: 9–17.
———. 1980b. *Quercus;* Keuringsrapport van de Keuringscommissie Sierbomen N.A.K.B. *Dendroflora* 17: 24–33.
Grounds, R. 1979. *Ornamental grasses.* New York: Van Nostrand Reinhold.
Grunert, C. 1982. *Gartenblumen von A bis Z.* Leipzig: Neumann Verlag.
Hachmann, H., P. Kiermeier, H. Reif, W. Schmalscheidt, H. Schwarz, H. D. Warda, and Chr. Zorn. 1986a. Gehölzsichtung. *Ergebnisse der sichtungsprüfungen bei Potentilla-niedrigwachsende arten—1980–1985.* Hannover: Bund deutscher Baumschulen (BdB) e.V. Pinneberg.
———. 1986b. Gehölzsichtung. *Ergebnisse der sichtungsprüfungen bei Pyracantha-niedrigwachsende arten—1980–1985.* Hannover:Bund deutscher Baumschulen (BdB) e.V. Pinneberg.
———. 1987. Gehölzsichtung. *Ergebnisse der sichtungsprüfungen bei Cotoneaster-niedrigwachsendearten—1980–1986.* Hannover:Bunddeutscher Baumschulen (BdB) e.V. Pinneberg.
Haegeman, J. 1978. *International list of tuberous begonia names.* Melle, Belgium: Rijksstation voor Sierplantenteelt.
———. 1979. *Tuberous begonias:* Origin and development. Vaduz, Liechtenstein: J. Cramer.

Hannibal, L. S. 1970–71. Garden crinums: An identification and checklist of crinums found in the United States. *Bull. Louisiana Soc. Hort. Res.* 3: 220–322.

Hansell, D. E., T. R. Dudley, and G. K. Eisenbeiss, eds. 1970. Handbook on hollies. *Am. Hort. Mag.* 49: 149–334.

Hansen, R., and J. Sieber. 1970. Die Sichtung des *Heuchera*–Sortimentes. *Deutsche Gärtnerbörse* 26: 635–637.

Harding, A. 1933. *Lilacs in my garden: a practical handbook for amateurs.* New York: Macmillan.

Haring, P. A., ed. 1986. *Modern roses 9.* Shreveport, Louisiana: Am. Rose Soc.

Harkness, M. G., and D. D'Angelo. 1986. *The Bernard E. Harkness seedlist handbook: a guide to the plants offered in the major plant societies' seed exchanges.* Portland, Oregon: Timber Press.

Harkness, P. 1991. *The photographic encyclopedia of roses.* New York: Gallery Books.

Harp, H. F. 1975. *Lythrums for home gardens.* Canad. Dept. Agric. Publ. 1285.

Harris, G. 1982. Japanese maples. *Plantsman* 3: 234–250.

Harris, J. G. S. 1983. An account of maples in cultivation. *Plantsman* 5: 35–58.

Harrison, C. R. 1975. *Ornamental conifers.* New York: Hafner Press.

Harvey, G. 1988. *Official nomenclature list.* Buderim, Queensland: Austral. Hibiscus Soc.

Haserodt, H., and T. D. Sydnor. 1983. Growth habits of five cultivars of *Gleditsia triacanthos. J. Arbor.* 9: 186–189.

Hashizume, T. c. 1982–1985. *Epiphyllums and other related genera.* Japan.

Hatch, L. 1982. *Nomenclatural analysis of the variegated hibernicas.* West Carrollton, Ohio: Am. Ivy Soc.

———. 1986. *Reference guide to the ornamental plant cultivars.* Raleigh, North Carolina: Taxonomic Computer Res.

Haworth-Booth, M. 1963. *The moutan or tree peony.* New York: St. Martin's Press.

———. 1984. *The hydrangeas.* 5th rev. ed. London: Constable.

Haysom, C. G. 1957. *Florists' auriculas and gold–laced polyanthus.* London: W. H. & L. Collingridge.

Hebb, R. S. 1971. The story of *Forsythia. Arnoldia* 31: 41–63.

Hecker, W. R. 1971. *Auriculas and primroses.* London: B. T. Batsford.

Heenan, P. B. 1991. *Checklist of Phormium cultivars.* Canterbury: Roy. New Zealand Inst. Hort.

Heieck, I. 1980. *Hedera sorten, ihre entstehung und geschichte dargestellt am sortiment der gärtnerei abtei neuburg.* Baiertal, Federal Republic of Germany: Garry Grüber.

Henkel, F., F. Rehnelt, and L. Dittmann. 1907. *Das buch der nymphaeaceen oder seerosengewächse.* Darmstadt: Friedrich Henkel.

Hensen, K. J. W. 1959. Het *Trollius*–sortiment. *Tuinbouwgids* 1959: 535–539.

———. 1963a. Identification of the Hostas ("Funkias") introduced and cultivated by von Siebold. *Meded. Landbouwhogeschool* 63(6): 1–22.

———. 1963b. Preliminary registration lists of cultivar names in *Hosta* Tratt. *Meded. Landbouwhogeschool* 63(7): 1–12.

———. 1966. Het in Nederland gekweekte Erigeronsortiment. *Beplantingen Boomkwekerij* 22: 131–134.

———. 1968. De in Nederland gekweekte Japanese anemonen. *Beplantingen Boomkwekerij* 24: 127–134.

———. 1969. Het in Nederland gekweekte *Astilbe* sortiment. *Beplantingen Boomkwekerij* 25: 106–112.

———. 1970 Het *Sorbus latifolia*-complex. *Meded. Bot. Tuinen Belmonte Arbor. Wageningen* 13: 181–194.

———. 1974. Het *Lavandula*-sortiment. *Groen* 30: 184–190.

———. 1976. Onderzoek en keuring van het *Aubrieta*-sortiment. *Groen* 32: 184–194.

———. 1979. Voortgezet onderzoek van het sortiment Japanse Anemonen. *Groen* 35: 363–370.

———. 1980. Taxonomie en nomenclatuur van het Vincasortiment. *Groen* 36: 187–194.

———. 1981. Het onderzoek van het sortiment Grasanjers (*Dianthus gratianopolitanus* en *plumarius*). *Groen* 37: 265–277.

———. 1983a. Some nomenclatural problems in *Hosta. Bull. Am. Hosta Soc.* 14: 41–45.

———. 1983b. Het onderzoek van het *Heliopsis* sortiment. *Groen* 39: 265–270.

———. 1985. A study of the taxonomy of cultivated hostas. *Plantsman* 7: 1–35.

Hensen, K. J. W., and N. Groendijk–Wilders. 1986a. An account of some sedums cultivated in Europe. *Plantsman* 8: 1–20.

———. 1986b. Siergrassen (*Gramineae, Cyperaceae en Juncaceae*). *Dendroflora* 23: 55–85.

Hertrich, W. 1954–1959. *Camellias in the Huntington Garden.* San Marino, California: Huntington Bot. Garden

Hillier, H. G. 1982. *The Hillier colour dictionary of trees and shrubs.* New York: Van Nostrand Reinhold.

———. 1991. *The Hillier manual of trees and shrubs.* 6th ed. London: David & Charles.

Hilton, R. J. 1982. Registration of *Amelanchier* cultivar names. *Fruit Var. J.* 36(4): 108–110.

———. 1984. *Amelanchier* cultivar addenda. *Fruit Var. J.* 38(4): 165.

Hodgkin, E. 1961. Daphnes. *J. Roy. Hort. Soc.* 86: 481–488.

Hodgson, R. W. 1967. Horticultural varieties of *Citrus*, p. 431–591. In: W. Reuther, H. J. Webber, and L. D. Batchelor (eds.). *The Citrus industry.* Vol. 1. Rev. ed. Berkeley: Univ. California Press.

Hogan, E. L., ed. 1988. *Sunset western garden book.* Menlo Park, California: Lane Publ.

Hoog, M. N. 1980. Bulbous irises. *Plantsman* 2: 141–164.

Hornback, B. 1994. New New Zealand flaxes. *Pacific Hort.* 55(3): 33–40.

Horobin, J. F. 1985. *Schlumbergera. Plantsman* 7: 53–63.

Horsman, J. 1984. Silver firs in cultivation. *Plantsman* 6: 65–100.

Hottes, A. C. 1916. *Gladiolus studies—II: varieties of the garden gladiolus.* Cornell Univ. Extens. Bull. 11.

Howard, R. A. 1961. Registration lists of cultivar names in *Cornus* L. *Arnoldia* 21: 9–18.

———. 1965. A check list of cultivar names in *Weigela. Arnoldia* 25: 49–69.

———. 1969. A check list of cultivar names used in the genus *Lantana. Arnoldia* 29: 73–109.

Howie, J. 1980. *Hibiscus: queen of the flowers. Nomenclature.* Brisbane, Australia: James Malcolm Howie.

Hu, Shiu-ying. 1954–1956. A monograph of the genus *Philadelphus. J. Arnold Arbor.* 35: 275–333; 36: 52–109, 325–368; 37: 15–90.

Hume, H. H. 1955. *Camellias in America.* Rev. ed. Harrisburg, Pennsylvania: J. Horace MacFarland.

———. 1961. The *Ophiopogon–Liriope* complex. *Baileya* 9: 134–158.

Hunt, P. F. 1986. The nomenclature and registration of orchid hybrids at specific and generic levels. Pp. 367–374 in B. T. Styles, ed., *Infraspecific classification of wild and cultivated plants*. Oxford: Clarendon Press.

Huttleston, D. G. 1986. Cultivar registration of unassigned woody ornamentals. *HortScience* 21: 361–366.

————. 1988. Cultivar registration of unassigned woody ornamentals. *HortScience* 23: 454–456.

————. 1989. International registrations of cultivar names for unassigned woody genera. *HortScience* 24: 430–432.

————. 1990. International registrations of cultivar names for unassigned woody genera. *HortScience* 25: 616–617.

————. 1991. International registrations of cultivar names for unassigned woody genera. *HortScience* 26: 475–476.

————. 1992. International registration of cultivar names for unassigned woody genera 1991. *HortScience* 27: 302.

————. 1993. International registration of cultivar names for unassigned woody genera 1992. *HortScience* 28: 278–279.

Hyatt, B. 1989. *Auriculas*. London: Cassell.

Ievinya, S. O., and M. A. Lusinya. 1975. *Astil'by*. Riga: Izdatal'stvo "Zinatne."

Ihei, I. 1984. A brocade pillow: *Azaleas of old Japan*. New York: Weatherhill.

Ilsink, L. K. J. 1988. Dendrologische notities. *Hydrangea paniculata* en de cultivars. *Groen* 44(10): 36–37.

Ilsink, L. K. J., and P. C. de Jong. 1986. Het geslacht *Euonymus*. II. De groenblijvende soorten. *Groen* 42(1): 23–27.

Ingles, J. 1990. *American Begonia Society listing of Begonia cultivars*. Rev. ed. Buxton check list. Am. Begonia Soc.

Ingram, C. 1948. *Ornamental cherries*. London: Country Life.

Ingram, J. 1963. Studies in the cultivated *Ericaceae*. 3. *Andromeda*. 4. *Pieris*. *Baileya* 11: 37–46.

International Poplar Commission. 1971. *Registration of poplar names*. Rome: Food & Agric. Org. United Nations FAO: CIP: 1/30.

————. 1990. *Catalogue international des cultivars de peupliers*. Rome: Food & Agric. Org. United Nations FOA: CIP: Catalogue/90.

Isaacson, R. T. 1989. *Andersen horticultural library's source list of plants and seeds*. Chanhassen, Minnesota: Andersen Hort. Libr.

Jacobsen, H. 1977. *Lexicon of succulent plants*. Poole, England: Blandford Press.

Jacobson, A. L. 1992. *Purpleleaf plums*. Portland, Oregon: Timber Press.

Jäger, A. 1960. *Rosenlexicon*. Leipzig: Zentral–Antiquariat Deutschen Demokratischen Republik.

Janaki Ammal, E. K. 1951. Chromosomes and evolution of the garden *Philadelphus*. *J. Roy. Hort. Soc.* 76: 269–275.

Japan Begonia Society. 1980. *Begonias* (in Japanese). Tokyo: Seibundo Shinkosha.

Japan Orchid Growers Association. [Not dated] *Quality stream cattleyas*. Japan: Josekishi.

Jaynes, R. A. 1975. *The laurel book*. New York: Hafner Press.

————. 1983. Checklist of cultivated laurels, *Kalmia* spp. *Bull. Am. Assoc. Bot. Gard.* 17(4): 99–106.

————. 1988. *Kalmia, the laurel book II*. Portland, Oregon: Timber Press.

————. 1989. *Kalmia* registration 1988. *HortScience* 24: 436–437.

Jaynes, R. A., and A. H. Graves. 1963. Connecticut hybrid chestnuts and their culture. *Connecticut Agric. Exp. Sta. Bull.* 657: 1–29.

Jaynes, R. A., A. J. Brand, and J. Arnow. 1993. Kousa dogwood. *Am. Nurseryman* 178(10): 40–47.

Jefferson, R. M. 1970. *History, progeny, and locations of crabapples of documented authentic origin.* Natl. Arbor. Contr. No. 2.

Jefferson, R. M., and K. K. Wain. 1984. The nomenclature of cultivated Japanese flowering cherries *(Prunus):* The Sato–sakura group. Natl. Arbor. Contr. No. 5.

Jelitto, L., and W. Schacht. 1990. *Hardy herbaceous perennials.* 2 vols. 3d ed. Trans. M. E. Epp. Portland, Oregon: Timber Press.

Jensma, J. R. 1989. *List of Aster cultivars 1989.* Wageningen: J. R. Jensma.

Jervis, R. N. 1980. *Aglaonema growers notebook.* Clearwater, Florida: Roy N. Jervis.

Johnson, A. T. 1956. *Hardy heaths.* London: Blandford Press.

Jones, D. L. 1987. *Encyclopedia of ferns.* Portland, Oregon: Timber Press.

Jong, P. C. de. 1986. *Betula:* Problematiek van de systematiek en de benaming. Betekenis en mogelijkheden voor de cultuur. *Dendroflora* 23: 3–18.

Kalva, V. 1980. *Cerini* (in Latvian). Riga: Liesma.

———. 1988. *Sirelid* (in Latvian). Riga: Vlagus.

Kapranova, N. N., and L. K. Lukina. 1972. Novye sorta *Philadelphus* L. selektsii lesostepnoi opytnoi stantsii. *Bjull. Glavn. Bot. Sada* 86: 81–87.

Kawakami, Mrs T. c. 1981. *The beautiful Saintpaulia* (in Japanese). Japan: Shufunotomo.

Keeley, J. E., and M. B. Keeley. 1994. Taxonomic affinities of *Arctostaphylos* and *Ceanothus* cultivars. *Fremontia* 22: 27–30.

Kelsey, H. P., and W. A. Dayton, eds. 1942. *Standardized plant names.* 2d ed. Harrisburg, Pennsylvania: J. Horace MacFarland.

Kessenich, G. M., ed. 1976. *Peonies: history of the peonies and their origination.* Hopkins, Minnesota: Am. Peony Soc.

———. 1990. *The American hybrid peony.* Hopkins, Minnesota: Am. Peony Soc.

Kharchenko, K. D. 1975. *Floksi* (in Russian). Kiev: Naukova Dumka.

Kington, S. 1989a. *The international daffodil checklist.* London: Roy. Hort. Soc.

———. 1989b. *The international daffodil register (1969).* 15th suppl. London: Roy. Hort. Soc.

———. 1990. *The international daffodil register (1969).* 16th suppl. London: Roy. Hort. Soc.

———. 1991. *The international daffodil register (1969).* 17th suppl. London: Roy. Hort. Soc.

———. 1992. *The international daffodil register (1969).* 18th suppl. London: Roy. Hort. Soc.

Kitchingham, R. M. 1985. Some species and cultivars of *Hemerocallis. Plantsman* 7: 68–89.

Knowlton, H., ed. 1959. *Iris check list of registered cultivar names 1950–1959.* St. Louis: Am. Iris Soc.

Köhlein, F. 1984. *Saxifrages and related genera.* Trans. D. Winstanley. London: B. T. Batsford.

Kortmann, J. P. 1987. *Cryptomeria japonica. Dendroflora* 24: 8–36.

Kostelijk, P. J. 1984. *Crocosmia* in gardens. *Plantsman* 5: 246–253.

Koster, R. 1972. Elf nieuwe populiereklonen: ten geleide. *Ned. Boschbouw–Tijdschr.* 44: 173–179.

Krauss, H. K. 1955. *Geraniums for home and garden.* New York: Macmillan.

Kraxberger, M., ed. 1980. *American Rhododendron hybrids.* Tigard, Oregon: Am. Rhododendron Soc.

Krüssmann, G. 1958a. Sichtung des Weigelien–Sortimentes. *Deutsche Baumschule* 10(1): 1–4.
——. 1958b. Das *Deutzia*–Sortiment. *Deutsche Baumschule* 10(8): 1–10.
——. 1958c. Das *Philadelphus*–Sortiment. *Deutsche Baumschule* 10(11): 1–10.
——. 1981. *The complete book of roses.* Trans. G. Krüssmann and N. Raban. Portland, Oregon: Timber Press.
——. 1984–1986. *Manual of cultivated broad–leaved trees & shrubs.* 3 vols. Trans. M. E. Epp. Portland, Oregon: Timber Press.
——. 1985. *Manual of cultivated conifers.* Trans. M. E. Epp. Portland, Oregon: Timber Press.
Krüssmann, G., W. Siebler, and W. Tangermann. 1970. *Winterharte Gardenstauden.* Berlin: Paul Parey.
Kudryavtseva, V. M. 1987. *Tyul'pany* (in Russian). Minsk: Polymya.
Kunming Institute of Botany, Academica Sinica. 1986. *Yunnan camellias of China.* Beijing: Science Press.
Kunst, A. G., and A. O. Tucker. 1989. "Where have all the flowers gone?" A preliminary list of origination lists for ornamental plants. *Assoc. Preservation Technol. Bull.* 21(2): 43–50.
Kusakabe, I. 1987. *Nepenthes* corrections to world carnivorous plant list. *Carnivorous Pl. Newslett.* 16: 102.
Laar, H. J. van de. 1966. *Pyracantha. Dendroflora* 3: 40–46.
——. 1967. *Hosta* (Funkia). *Dendroflora* 4: 30–41.
——. 1968. *Calluna vulgaris. Dendroflora* 5: 11–12.
——. 1969. *Pernettya mucronata. Dendroflora* 6: 53–56.
——. 1970a. *Calluna* en *Erica. Dendroflora* 7: 6–32.
——. 1970b. *Prunus laurocerasus. Dendroflora* 7: 42–61.
——. 1971. *Cytisus* en *Genista. Dendroflora* 8: 3–18.
——. 1972. *Berberis. Dendroflora* 9: 9–37.
——. 1973. *Bergenia. Dendroflora* 10: 3–14.
——. 1974. *The heather garden.* London: Collins.
——. 1975. *Mahonia: Mahonia* en *Mahoberberis. Dendroflora* 11/12: 18–35.
——. 1977a. *Calluna* en *Erica* (geelbladige cultivars). *Dendroflora* 13/14: 17–33.
——. 1977b. *Daboecia. Dendroflora* 13/14: 34–39.
——. 1979. *Euonymus*—bladhoudende soorten en cultivars. *Dendroflora* 15/16: 9–23.
——. 1981a. *Epimedium. Dendroflora* 18: 5–13.
——. 1981b. *Parthenocissus tricuspidata. Dendroflora* 18: 41–50.
——. 1982. *Potentilla fruticosa. Dendroflora* 19: 29–44.
——. 1984. *Skimmia. Dendroflora* 21: 63–80.
——. 1985. *Clematis;* grootbloemige hybriden. *Dendroflora* 22: 33–58.
——. 1988. *Lonicera* (Klimmende soorten, varièteiten en cultivars). *Dendroflora* 25: 36–54.
——. 1989. *Naamlijst van houtige gewassen.* Boskoop: Proefstation voor de Boomwekerij.
——. 1992. *Parthenocissus*—Sortiment: Durcheinander fördert das Geschäft. *Deutsche Baumschule* 12: 608–609.
Laar, H. J. van de, and G. Fortgens. 1990. *Naamlijst van vaste planten.* Boskoop: Proefstation voor de Boomkwekerij.
Lancaster, R. 1970. Complete guide to *Hamamelis*—the witch hazels. *Gard. Chron.* 167(23): 26–29, (24): 24–27.
——. 1981. An account of *Euonymus* in cultivation and its availability in commerce. *Plantsman* 3: 133–166.
Langdon, B. 1969. *The tuberous Begonia.* London: Cassell.

Larkman, B. 1985. Australia's glorious grevilleas. *Am. Nurseryman* 106(9): 106–108, 110, 112, 116, 118–119.

Lawrence, G. H. M. 1956. The cultivated ivies. *Morris Arbor. Bull.* 7: 19–31.

Lawrence, G. H. M., and A. E. Schulze. 1942. The cultivated hederas. *Gentes Herb.* 6: 106–173.

Lawson, A. H. 1968. *Bamboos.* New York: Taplinger Publ.

Leach, D. G. 1961. *Rhododendrons of the world and how to grow them.* New York: Charles Scribner's Sons.

Lee, F. P., F. O. Coe, B. Y. Morrison, M. Perkins, and F. Weiss. 1965. *The azalea book.* New York: Van Nostrand.

Lee, G. S., ed. 1966. Daffodil handbook. *Am. Hort. Mag.* 45: 1–227.

Leeuwenberg, A. J. M. 1979. The *Loganiaceae* of Africa XXVII. *Buddleja* L. II. Revision of the African and Asiatic species. *Meded. Landbouwhogeschool* 79(6): 1–163.

Lemmens, R. H. M. J. 1985. *Desmodium, Hedysarum, Indigofera* en *Lespedeza. Dendroflora* 22: 59–61.

Leslie, A. C. 1982. *The international lily register.* London: Roy. Hort. Soc.

———. 1986. International plant registration. Pp. 357–365 in B. T. Styles, ed., *Infraspecific classification of wild and cultivated plants.* Oxford: Clarendon Press.

Letts, J. F. 1966. *Handbook of hardy heaths and heathers: hardy, free-flowering, foliage, evergreen plants.* Windlesham, England: John F. Letts.

Leue, M. c. 1987. *Epiphyllum: the splendor of leaf cacti.* Haunetal: Marga Leue.

Lewis, J. 1985. *The international conifer register. A preliminary list comprising registrations of cultivars of the Coniferopsida 1947–1984.* London: Roy. Hort. Soc.

———. 1986a. The classification of conifer cultivars. *Acta Hort.* 182: 159–175.

———. 1986b. Conifer registration and the conifer register. *Acta Hort.* 182: 407–410.

Lewis, J., and A. C. Leslie. 1987. *The international conifer register. Pt. 1—Abies to Austrotaxus.* London: Roy. Hort. Soc.

———. 1989. *The international conifer register. Pt. 2—Belis to Pherosphaera (excluding the cypresses and Juniperus).* London: Roy. Hort. Soc.

Lewis, J., and M. Lynch. 1989. *Campanulas.* Portland, Oregon: Timber Press.

Lindgren, D. T. 1993. The need to register new cultivar names: *Penstemon,* a case study. *HortScience* 28: 82–83.

Lindgren, D. T., and B. Davenport. 1992. List and descripton of named cultivars in the genus *Penstemon* (1992). Univ. Nebraska Coop. Ext. EC 92–1246–D.

Livingston, P. A., and F. H. West, eds. 1978. *Hybrids and hybridizers: rhododendrons and azaleas for eastern north america.* Newton Square, Pennsylvania: Harrowwood Books.

Lloyd, C. 1965. *Clematis.* London: Country Life.

———. 1989. *Clematis.* Deer Park, Wisconsin: Capability's Books.

Loewer, P., ed. 1988. Ornamental grasses. *Pl. & Gard.* 44(3): 1–104.

Lombarts, P. 1984. *Malus*—sierappels; Keuringsrapport, "Technische Keuringscommissie—Houtige Siergewassen" van de N.A.K.B. *Dendroflora* 21: 39–62.

Lord, T. 1988. *Aconitum*—notes on the genus. Hardy Pl. Soc. Bull. 10(2): 85–91.

Luneva, Z. S., N. L. Mikhailov, and E. A. Sudakova. 1989. *Siren'* (in Russian). Moscow: Agropromiedat.

Lyall, H. G. 1959. *Hardy primulas.* London: W. H. & L. Collingridge.

Maatsch, R. 1958. *Vorläufige sortenliste von Callistephus chinensis Nees.* Hannover: Inst. Zierpflanzenbau Techn. Hochschule.

——. 1962. Registrierung des Sortimentes der Beetbegonien (*Begonia* ×*semperflorens*-Cultorum) von 1900–1961. *Gartenbauwissenschaft* 27: 399–412.

——. 1964. *Sortenliste von Callistephus chinensis Nees*. 2. Folge. Hannover: Inst. Zierpflanzenbau Techn. Hochschule.

——. 1980. *Das buch der freilandfarne*. Berlin: Paul Parey.

Maatsch, R., and G. Nolting. 1968. Registrierung des Sortimentes von *Petunia* ×*hybrida* Vilm. *Gartenbauwissenschaft* 33: 285–316.

——. 1969. Registrierung des Sortimentes der Beetbegonien (*Begonia* ×*semperflorens* Cultorum) II. Nachtrag von 1961–1968. *Gartenbauwissenschaft* 34: 281–285.

——. 1970. *Sortenliste von Tagetes erecta L., Tagetes patula L. und Tagetes tenuifolia Cav. (Internationale Registrierliste)*. Hannover: Inst. Zierpflanzenbau Techn. Univ.

——. 1971a. Registrierung des Sortiments der Beetbegonien (*Begonia* ×*semperflorens* cultorum). III. 2. Nachtrag von 1968–1971. *Gartenbauwissenschaft* 36: 201–204.

——. 1971b. Registrierung des Sortiments von *Petunia* ×*hybrida* Vilm. 1. Nachtrag von 1968–1971. *Gartenbauwissenschaft* 36: 241–249.

——. 1971c. *Sortenliste von Callistephus chinensis Nees (Internationale Registerliste). 3. Folge.* Hannover: Inst. Zierpflanzenbau Techn. Univ.

Mabberly, D. J. 1984. A monograph of *Melia* in Asia and the Pacific: The history of white cedar and Persian lilac. *Gard. Bull. Straits Settlem.* 37: 49–64.

MacDaniels, L. H. 1981. A study of cultivars in *Bougainvillea* (*Nyctaginaceae*). *Baileya* 21: 77–100.

Macoboy, S. 1981. *The colour dictionary of camellias.* Sydney: Lansdowne Press.

Mallet, C., R. Mallet, and H. van Trier. 1992. *Hydrangeas: species and cultivars.* Varengeville s/mer, France: Centre d'Art Floral.

Mallett, A. 1983. Elementary "geumetry." *Bull. Hardy Pl. Soc.* 6(3): 95–98.

Mansfield, T. C. 1943. *Roses in colour and cultivation.* London: William Collins.

——. 1951. *Carnations in colour and cultivation.* London: Collins.

Manthey, G. 1990. *Fuchsias.* Trans. D. Christie. Portland, Oregon: Timber Press.

Markham, E. 1935. *The large & small flowered Clematis and their cultivation in the open air.* New York: Charles Scribner's Sons.

Mast, H. 1986. The naming of plants under the UPOV Convention. Pp. 399–417 in B. T. Styles, ed., *Infraspecific classification of wild and cultivated plants.* Oxford: Clarendon Press.

Mathew, B. 1982. *Pulmonaria* in gardens. *Plantsman* 4: 100–111.

——. 1989a. *The genus Lewisia.* Portland, Oregon: Timber Press.

——. 1989b. *Hellebores.* Alpine Garden Soc., St. John's Woking, England.

Matthews, L. J. 1983. *South African Proteaceae in New Zealand.* Manakau via Levin, New Zealand: Matthews Publ.

McArdle, A. J., and F. S. Santamour. 1984. Checklists of cultivars of European ash (*Fraxinus*) species. *J. Arbor.* 10: 21–32.

——. 1985. Cultivar checklist for English oak (*Quercus robur* L.). *J. Arbor.* 11: 307–315.

——. 1987a. Cultivar checklist of white oak species (excl. *Quercus robur* L.). *J. Arbor.* 13: 203–208.

——. 1987b. Cultivar checklist of *Quercus* (excl. subg. *Quercus*). *J. Arbor.* 13: 250–256.

McClintock, D. 1986. Harmonising botanical and cultivar classification with special reference to hardy heathers. *Acta Hort.* 182: 277–283.

McClintock, E. 1982. Erythrinas cultivated in California. *Allertonia* 3: 139–154.

McKelvey, D. 1928. *The lilac*. New York: Macmillan.

McWilliam, A. 1973. Official list of named varieties and hybrids registered by the American Penstemon Society: 1958–1973. *Bull. Am. Penstemon Soc.* 32: 7B–7C.

——. 1977. Report of the registrar of named varieties and hybrids. *Bull. Am. Penstemon Soc.* 36: 7.

Median Iris Society. 1984. *The median bearded irises: introduction and varietal listing*. Wichita, Kansas: Median Iris Soc.

——. 1992. *Median bearded irises: introduction and varietal listing through 1990*. Tipp City, Ohio: Median Iris Soc.

Meier, W. 1973a. *Das Aster amellus-sortiment*. Oeschberg, Switzerland: W. Meier.

——. 1973b. *Das Aster dumosus-sortiment*. Oeschberg, Switzerland: W. Meier.

——. 1973c. *Das Aster novae-angliae-sortiment*. Oeschberg, Switzerland: W. Meier.

——. 1973d. *Das Aster novi-belgii-sortiment*. Oeschberg, Switzerland: W. Meier.

Meikle, C. E., ed. 1980. *Modern roses 8*. Harrisburg, Pennsylvania: J. Horace MacFarland.

Mellichamp, L., and R. Gardner. 1987. New cultivars of *Sarracenia*. *Carnivorous Pl. Newslett.* 16: 39–42.

Menninger, E. D. 1960. Catalog of hybrid *Nerine* clones 1882–Dec. 31, 1958. *Pl. Life* 16: 63–74.

Metcalf, L. J. 1963. Check list of *Leptospermum* cultivars. *J. Roy. New Zealand Inst. Hort.* 5: 224–230.

Meyer, F. 1952. *Flieder*. Stuttgart: Eugen Ulmer.

Meyer, M. H. 1975. *Ornamental grasses: decorative plants for home and garden*. New York: Charles Scribner's Sons.

Mickel, J. T. 1994. *Ferns for American gardens*. New York: Macmillan.

Misono, I. c. 1974–1978. *Begonias*. 2 vols. Trans. A. M. DeCola and H. Arakawa. Los Angeles: Am. Begonia Soc.

Mitchell, P. J. c. 1973. *The Sempervivum & Jovibara handbook*. Burgess Hill, England: Sempervivum Soc.

——. 1982. Tentative international register of cultivars of *Jovibara, Rosularia* and *Sempervivum*, Pt. 1. *Houselekes* 13: 64–86.

——. 1983. Tentative international register of cultivars of *Jovibara, Rosularia* and *Sempervivum*, Pt. 2. *Houselekes* 14: 4–40.

——. 1985. *International cultivar register for Jovibarba, Rosularia, Sempervivum. Vol. 1*. Burgess Hill, England: Sempervivum Soc.

Miyoshi, M. 1916. Japanische Bergkirschen, ihre Wildformen und Kulturrassen. Ein Beitrag zur Formenlehre. *J. Coll. Sci. Imp. Univ. Tokyo* 24: 1–175.

Moody, M., ed. 1992. *The illustrated encyclopedia of roses*. Portland, Oregon: Timber Press.

Moore, H. E. 1953. Comments on some kohlerias cultivated in the United States. *Baileya* 1: 89–101.

——. 1955a. Pelargoniums in cultivation. I. *Baileya* 3: 5–25, 41–46.

——. 1955b. Pelargoniums in cultivation. II. *Baileya* 3: 70–97.

Morgenstern, K. D. 1979. *Sansevierias*. Kempten, Germany: Illertaler Offsetdrucken & Verlag GmbH

Morrison, B. Y. 1953. *The Glenn Dale azaleas*. Agric. Monogr. U.S.D.A. 20.

Muir, N. 1984. A survey of the genus *Tilia*. *Plantsman* 5: 206–242.

——. 1988. Additional notes on hybrid limes *(Tilia)*. *Plantsman* 10: 104–127.

Mukherjee, I., and T. N. Khoshoo. 1970. Genetic–evolutionary studies on cultivated cannas. VII: Taxonomic treatment and horticultural classification. *J. Bombay Nat. Hist. Soc.* 67: 390–397.

Mulligan, B. O. 1958. *Maples cultivated in the United States and Canada.* Am. Assoc. Bot. Gard.

Munson, R. H. 1973. Vegetative key to selected dwarf and slow–growing conifers. M.S. Thesis. Ithaca: Cornell Univ.

——. 1981. Integrated methods of cultivar identification: A case study of selected *Ericaceae.* Ph.D. Thesis. Ithaca: Cornell Univ.

——. 1984. Heaths and heathers cultivated in North America (*Ericaceae*). *Baileya* 22: 101–133.

——. 1989. *Hemerocallis, the daylily.* Portland, Oregon: Timber Press.

Murray, A. E. 1970. A monograph of the *Aceraceae.* Ph.D. Thesis. Philadelphia: Univ. Penn.

Müssel, H. 1971. Das *Sorbus*–Sortiment in Weihenstephan unter besonderer Berücksichtigung der Lombarts–Hybriden. Pp. 1–10 in Jahresbericht 1971. Weihenstephan: Germany: Fachhochschule.

——. 1986. A study on the cultivars of the *Aconitum napellus* and *A. variegatum* complex according to the characteristics for destination of the inferior taxa. *Acta Hort.* 182: 89–94.

Nakamura, M. 1964. *Adonis amurensis* in Japan. *J. Roy. Hort. Soc.* 89: 121–125.

Nannenga–Bremekamp, N. E. 1970. Notes on *Hedera* species, varieties and cultivars grown in the Netherlands. *Meded. Bot. Tuinen Belmonte Arbor. Wageningen* 13: 195–212.

Narain, P. 1988. *Gloriosa:* Cultivars and natural species. *Herbertia* 44: 2–12.

National Chrysanthemum Society. 1991. *Handbook on chrysanthemum classification.* Natl. Chrysanthemum Soc.

Nelson, K., ed. 1971. *Iris check list of registered cultivar names 1960–1969.* Tulsa, Oklahoma: Am. Iris Soc.

Nelson, K., and K. Keppel, eds. 1981. *Iris check list of registered cultivar names 1970–1979.* Tulsa, Oklahoma: Am. Iris Soc.

——, eds. 1991. *Registrations and introductions in 1990.* Tulsa, Oklahoma: Am. Iris Soc.

——, eds. 1992a. *Iris check list of registered cultivar names 1980–1989.* Tulsa, Oklahoma: Am. Iris Soc.

——, eds. 1992b. *Registrations and introductions in 1991.* Tulsa, Oklahoma: Am. Iris Soc.

——, eds. 1993. *Registrations and introductions in 1992.* Tulsa, Oklahoma: Am. Iris Soc.

Newsholme, C. 1992. *Willows: the genus Salix.* Portland, Oregon: Timber Press.

Nienstaedt, H., and A. H. Graves. 1955. Blight resistant chestnuts. *Connecticut Agric. Exp. Sta. Circ.* 192: 1–18.

Nietner, T. E. 1880. *Die Rose.* Berlin: Wiegandt, Hempel, & Parey.

Niwa, T. 1936. *Chrysanthemums of Japan.* Tokyo: Sanseido.

Nolting, G., and K. Zimmer. 1975a. Registrierung des Sortiments der Beetbegonien (*Begonia* ×*semperflorens* Cultorum). IV. 3. Nachtrag von 1971–1975. *Gartenbauwissenschaft* 40: 188–190.

——. 1975b. Registrierung des Sortiments von *Petunia* ×*hybrida* Vilm. III. 2. Nachtrag von 1971–1975. *Gartenbauwissenschaft* 40: 234–239.

——. 1975c. *Sortenlisten von Callistephus chinensis Nees (Internationale Registerliste). 4. Folge. Tagetes erecta L., Tagetes patula L. und Tagetes tenuifolia Cav. (Internationale Registerliste). 2. Folge.* Hannover: Inst. Zierpflanzenbau Techn. Univ.

——. 1980a. Registrierung des Sortiments der Beetbegonien (*Begonia Semperflorens*-Hybriden). V. 4. Nachtrag von 1975–1979. *Gartenbauwissenschaft* 45: 42–45.

——. 1980b. Registrierung des Sortiments von *Petunia* ×*hybrida* Vilm. IV. 3. Nachtrag von 1975–1979. *Gartenbauwissenschaft* 45: 139–143.

——. 1981. *Sortenlisten von Callistephus chinensis Nees (Internationale Registerliste). 5. Folge. Tagetes erecta L., Tagetes patula L. und Tagetes tenuifolia Cav. (Internationale Registerliste). 3. Folge.* Hannover: Inst. Zierpflanzenbau Techn. Univ.

——. 1984. Registrierung des Sortiments von *Petunia* ×*hybrida* Vilm. V. 4. Nachtrag von 1979–1983. *Gartenbauwissenschaft* 49: 280–283.

——. 1985. Registrierung des Sortiments der Beetbegonien (*Begonia* ×*semperflorens* Cultorum). VI. 5. Nachtrag von 1979–1983. *Gartenbauwissenschaft* 50: 93–95.

——. 1987. *Sortenlisten von Begonia-Semperflorens-Hybriden. VII. 6. Nachtrag von 1983–1987. Callistephus chinensis Nees (Internationale Registerliste). 6. Folge von 1981–1987. Petunia ×hybrida Vilm. (Internationale Registerliste). VI. 5. Nachtrag von 1983–1987. Tagetes erecta L., Tagetes patula L. und Tagetes tenuifolia Cav. (Internationale Registerliste). 4. Folge von 1981–1987.* Hannover: Techn. Inst. Zierpflanzenbau Univ.

1924. *Seven thousand dahlias in cultivation.* College Park, Maryland: J. B. S. Norton.

Norton, J. B. S., W. F. Stuntz, and W. R. Ballard. 1949. *Descriptive catalog of Hemerocallis clones 1893 to 1948.* Stanford, California: Am. Pl. Life Soc.

Obrizok, B. 1991. *Dwarf conifer selection guide and checklist.* Lagrangeville, New York: Dutchess County Conifers.

Ogg, S. 1961. *Delphiniums for everyone.* London: Blandford Press.

Ohwi, J., and Y. Ohta. 1973. *Flowering cherries of Japan.* Tokyo: Heibonsha.

Olmsted, F. L., F. V. Coville, and H. P. Kelsey. 1923. *Standardized plant names.* Salem, Massachusetts: Am. Joint Committee Hort. Nomencl.

Ottesen, C. 1989. *Ornamental grasses: the amber wave.* New York: McGraw-Hill.

Oudolf, P. 1993. Neue *Monarda*-Sorten aus Holland. *Gartenpraxis* (2): 8–16.

Oudshoorn, W. 1975. Vaste planten terzijde. Aantekeningen t.a.v. het *Erigeron*-sortiment. *Groen* 31(6): 180.

Pagen, F. J. J. 1987. *Oleanders: Nerium L. and the oleander cultivars.* Agric. Univ. Wageningen Pap. 87-2.

Park, B. 1956. *Collins guide to roses.* London: Collins.

Parker, P. F. 1986. A computerised data base for the identification and classification of cultivars of *Fuchsia* ×*hybrida* Voss. *Acta Hort.* 182: 411–416.

Paul, W. 1863. Notes on the varieties of English holly. *Proc. Roy. Hort. Soc.* 3: 110–117.

Peckham, Mrs W. H., ed. 1929. *Alphabetical Iris check list.* Baltimore: Am. Iris Soc.

Peckham, E. A. S., ed. 1940. *Alphabetical Iris check list.* Baltimore: Am. Iris Soc.

Pedley, R., and K. Pedley. 1974. *Coleus: a guide to cultivation and identification.* Edinburgh: John Bartholomew and Son.

Philip, C. 1992. *The plant finder. 6th ed.* Ed. T. Lord. Whitbourne, England: Hardy Pl. Soc.

Phillips, C. E. L., and P. N. Barber. 1967. *The Rothschild rhododendrons: a record of the gardens at Exbury.* London: Cassell.

————. 1979. *The Rothschild rhododendrons: a record of the gardens at Exbury*. Rev. ed. New York: Macmillan.

Phillips, R., and M. Rix. 1988. *Roses*. New York: Random House.

————. *The Random House book of perennials*. 2 vols. New York: Random House.

Pierot, S. W. 1974. *The ivy book*. New York: Macmillan.

Pieters, A. J. 1905. *A variety collection of Gladiolus*. U.S.D.A. Bur. Pl. Industr. Bull. 177.

Pietropaolo, J., and P. Pietropaolo. 1986. *Carnivorous plants of the world*. Portland, Oregon: Timber Press.

Plumeria Society of America. 1988. *Register of Plumeria cultivars*. 2d ed. Plumeria Soc. Am.

Poliakoff, A. T. 1987. *Guide to orchid hybrids: the Vanda family*. Alexandria, Virginia: Potomac Data Press.

Praeger, R. L. 1921. An account of the genus *Sedum* as found in cultivation. *J. Roy. Hort. Soc.* 46: 1–314.

Preston, I. 1944. Rosybloom crabapples for northern gardens. *J. New York Bot. Gard.* 45: 169–174.

Pridham, A. M. S. 1932. *The gladiolus: its history, classification, and culture*. Cornell Univ. Agric. Exp. Sta. Bull. 231.

Pringle, J. S. 1973. The cultivated taxa of *Clematis*, Sect. *Atragene (Ranunculaceae)*. *Baileya* 19: 49–89.

Proudley, B., and V. Proudley. 1974. *Heathers in color*. New York: Hippocrene Books.

————. 1975. *Fuchsias in color*. New York: Hippocrene Books.

Puttock, A. G. 1957. *Primulas*. London: John G. Gifford.

Rainbow Gardens. 1979. *Directory of epiphyllums and other related epiphytes*. Vista, California: Rainbow Gardens.

Ranson, E. R. 1946. *Michaelmas daisies and other garden asters*. London: John G. Gifford.

Raulston, J. C. 1984. Plants in the NCSU Arboretum—*Nandina domestica*—"heavenly bamboo". *Friends North Carolina State Arbor. Newslett.* 11: 5–9.

————. 1990. Redbud. *Am. Nurseryman* 171(5): 39–51.

————. 1992. Styrax. *Am. Nurseryman* 176(9): 23–32, 34.

Rector, A. 1963. *The master list of african violets . . . 1935 to 1963*. Ed. A. Wright. Knoxville: African Violet Soc. Am.

Rehder, A. 1940. *Manual of cultivated trees and shrubs hardy in North America*. 2d ed. New York: Macmillan.

————. 1949. *Bibliography of cultivated trees and shrubs hardy in the cooler temperate regions of the northern hemisphere*. Jamaica Plain, Massachussets: Arnold Arbor.

Reinhardt, T. A., M. Reinhardt, and M. Mopskowitz. 1989. *Ornamental grass gardening*. Los Angeles: HP Books.

Rhodes, H. L. J. 1954. The cultivated shrubby potentillas. *Baileya* 2: 89–96.

Rice, G., and E. Strangman. 1993. *The gardener's guide to growing hellebores*. Portland, Oregon: Timber Press.

Roberts, A. V. 1984. The chromosomes of *Nerine*. *The Garden* (London) 109: 413–415.

Robinson, J. T. 1989. New CP cultivar received in 1988. *Carnivorous Pl. Newslett.* 18: 39.

Rogers, O. M. 1976. *Tentative international register of cultivar names in the genus Syringa*. New Hampshire Agric. Exp. Sta. Res. Rep. 49.

Rogerson, B. 1985. *Clematis viticella* and its progeny. *Pacific Hort.* 46(1): 40–44.

Roller, K. J., D. H. Thibault, and V. Hidahl. 1972. *Guide to the identification of poplar cultivars on the prairies.* Canad. For. Serv. Publ. 1311.

Rose, P. Q. 1980. *Ivies.* Poole, England: Blandford Press.

Royal General Bulbgrowers' Association. 1971. *Classified list and international register of tulip names.* Hillegom, Netherlands: Koninklijke Algemeene Vereeniging voor Bloembollencultuur.

———. 1976. *Classified list and international register of tulip names.* Hillegom, Netherlands: Koninklijke Algemeene Vereeniging voor Bloembollencultuur.

———. 1981. *Classified list and international register of tulip names.* Hillegom, Netherlands: Koninklijke Algemeene Vereeniging voor Bloembollencultuur.

———. 1991. *International checklist for hyacinths and miscellaneous bulbs.* Hillegom, Netherlands: Koninklijke Algemeene Vereeniging voor Bloembollencultuur.

Royal General Bulbgrowers' Society. 1958. *A classified list of tulip names.* Haarlem, Netherlands: Koninklijke Algemeene Vereeniging voor Bloembollencultuur.

———. 1960. *Classified list and international register of tulip names.* Haarlem, Netherlands: Koninklijke Algemeene Vereeniging voor Bloembollencultuur.

———. 1965. *Classified list and international register of tulip names.* Haarlem, Netherlands: Koninklijke Algemeene Vereeniging voor Bloembollencultuur.

———. 1969. *Classified list and international register of tulip names.* Haarlem, Netherlands: Koninklijke Algemeene Vereeniging voor Bloembollencultuur.

Royal Horticultural Society. 1902. Report on perennial asters grown at Chiswick, 1902. *J. Roy. Hort. Soc.* 27: 638–648.

———. 1908a. Asters at Wisley, 1906–07. *J. Roy. Hort. Soc.* 33: 184–211.

———. 1908b. Cannas at Wisley, 1906–07. *J. Roy. Hort. Soc.* 33: 212–222.

———. 1908c. Tulips at Wisley, 1906–07. *J. Roy. Hort. Soc.* 33: 232–280.

———. 1908d. Classified list of daffodil names. London: Roy. Hort. Soc.

———. 1909. Cannas out of doors at Wisley, 1908. *J. Roy. Hort. Soc.* 34: 299–302.

———. 1910. Classified list of daffodil names. London: Roy. Hort. Soc.

———. 1912. Violas at Wisley, 1912. *J. Roy. Hort. Soc.* 38: 275–287.

———. 1913a. Violas at Wisley, 1913. *J. Roy. Hort. Soc.* 39: 381–400.

———. 1913b. Antirrhinums at Wisley, 1913. *J. Roy. Hort. Soc.* 39: 635–656.

———. 1917. Report of the tulip nomenclature committee 1914–1915. London: Roy. Hort. Soc.

———. 1926a. Perennial asters at Wisley, 1925. *J. Roy. Hort. Soc.* 51: 101–108.

———. 1926b. Sweet peas at Wisley, 1921–25. *J. Roy. Hort. Soc.* 51: 109–112.

———. 1926c. Annual scabious at Wisley, 1924. *J. Roy. Hort. Soc.* 51: 113–118.

———. 1926d. Helianthemums at Wisley, 1924–25. *J. Roy. Hort. Soc.* 51: 119–123.

———. 1926e. Delphiniums at Wisley, 1924–25. *J. Roy. Hort. Soc.* 51: 124–137.

———. 1928. Bearded irises tried at Wisley, 1925–27. *J. Roy. Hort. Soc.* 53: 116–160.

———. 1929. *A tentative list of tulip names.* London: Roy. Hort. Soc.

———. 1930a. *A tentative list of tulip names.* Suppl. no. 1. London: Roy. Hort. Soc.

———. 1930b. Bearded irises tried at Wisley, 1928–29. *J. Roy. Hort. Soc.* 55: 132–140.

———. 1930c. Gaillardias at Wisley, 1929. *J. Roy. Hort. Soc.* 5: 141–144.

———. 1931a. Perennial lupins tried at Wisley, 1929–30. *J. Roy. Hort. Soc.* 56: 115–120.

———. 1931b. *Classified list of daffodil names.* London: Roy. Hort. Soc.

———. 1938. *Classified list of daffodil names.* London: Roy. Hort. Soc.

———. 1939. *Classified list of tulip names.* London: Roy. Hort. Soc.

———. 1948. *Classified list of daffodil names.* London: Roy. Hort. Soc.

———. 1955. *Classified list of daffodil names.* London: Roy. Hort. Soc.

——. 1958. *Classified list of daffodil names.* London: Roy. Hort. Soc.

——. 1961. *Classified list of daffodil names.* London: Roy. Hort. Soc.

——. 1964. *Rhododendron handbook 1964. pt. 2. Rhododendron hybrids.* London: Roy. Hort. Soc.

——. 1965. *Classified list of daffodil names.* London: Roy. Hort. Soc.

——. 1969a. *Tentative classified list and international register of Dahlia names 1969.* London: Roy. Hort. Soc.

——. 1969b. *Rhododendron handbook 1969. pt. 2. Rhododendron hybrids.* London: Roy. Hort. Soc.

——. 1969c. *Classified list of daffodil names.* London: Roy. Hort. Soc.

——. 1970a. *A tentative check-list of Delphinium names.* London: Roy. Hort. Soc.

——. 1970b. Astilbes. *Proc. Roy. Hort. Soc. London* 95: 132–135.

——. 1972. *Sander's list of orchid hybrids. add. 1961–1970: containing the names and parentage of all orchid hybrids registered from 1st January 1961 to 31st December 1970.* London: Roy. Hort. Soc.

——. 1974. *International Dianthus register.* London: Roy. Hort. Soc.

——. 1975. *Classified list and international register of daffodil names 1960–1975.* London: Roy. Hort. Soc.

——. 1978. *Agapanthus. Proc. Roy. Hort. Soc. London* 103: 89–93.

——. 1980. *Sander's list of orchid hybrids. add. 1971–1975: containing the names and parentages of all orchid hybrids registered from 1st January 1971 to 31st December 1975.* London: Roy. Hort. Soc.

——. 1981. *Sander's list of orchid hybrids. add. 1976–1980: containing the names and parentages of all orchid hybrids registered from 1st January 1976 to 31st December 1980.* London: Roy. Hort. Soc.

——. 1982. *First supplement to the international lily register.* London: Roy. Hort. Soc.

——. 1983. *The international Dianthus register. 2d ed.* London: Roy. Hort. Soc.

——. 1984a. *Second supplement to the international lily register 1982.* London: Roy. Hort. Soc.

——. 1984b. *The international Dianthus register. 2d ed. (1983). 1st suppl.* London: Roy. Hort. Soc.

——. 1985a. *Third supplement to the international lily register 1982.* London: Roy. Hort. Soc.

——. 1985b. *The international Dianthus register. 2d ed. (1983). 2nd suppl.* London: Roy. Hort. Soc.

——. 1985c. *Handbook on orchid nomenclature and registration. 3rd ed.* London: Roy. Hort. Soc.

——. 1986a. *Fourth supplement to the international lily register 1982.* London: Roy. Hort. Soc.

——. 1986b. *The international Dianthus register. 2d ed. (1983). 3rd suppl.* London: Roy. Hort. Soc.

——. 1986c. *Sander's list of orchid hybrids. add. 1981–1985: containing the name and parentage of all orchid hybrids registered from 1st January 1981 to 31st December 1985.* London: Roy. Hort. Soc.

——. 1987a. *Fifth supplement to the international lily register 1982.* London: Roy. Hort. Soc.

——. 1988a. *The international lily register (1982). 6th suppl.* London: Roy. Hort. Soc.

——. 1988b. *The international Dianthus register (1983). 4th suppl.* London: Roy. Hort. Soc.

————. 1988c. *The international Dianthus register (1983)*. *5th suppl*. London: Roy. Hort. Soc.

————. 1988d. *The International Dahlia register (1969)*. *2nd suppl*. London: Roy. Hort. Soc.

————. 1988e. *The international Rhododendron register (1958)*. *28th suppl*. London: Roy. Hort. Soc.

————. 1989a. *The international lily register (1989)* *7th suppl*. London: Roy. Hort. Soc.

————. 1989b. *The international Dianthus register (1983)*. *6th suppl*. London: Roy. Hort. Soc.

————. 1989c. *The international Dahlia register (1969)*. *3d suppl*. London: Roy. Hort. Soc.

————. 1989d. *The international Rhododendron register: checklist of rhododendron names registered 1959–1987*. London: Roy. Hort. Soc.

————. 1989e. *The international Rhododendron register (1958)*. *29th suppl*. London: Roy. Hort. Soc.

————. 1990a. *The international Dianthus register (1983)*. *7th suppl*. London: Roy. Hort. Soc.

————. 1990b. *The international lily register (1982)*. *8th suppl*. London: Roy. Hort. Soc.

————. 1990c. *The international Rhododendron register (1958)*. *30th suppl*. London: Roy. Hort. Soc.

————. 1991a. *The international lily register (1982)*. *9th suppl*. London: Roy. Hort. Soc.

————. 1991b. *The international Rhododendron register (1958)*. *31st suppl*. London: Roy. Hort. Soc.

————. 1992a. *The international Dahlia register (1969)*. *4th suppl*. London: Roy. Hort. Soc.

————. 1992b. *The international lily register (1982)*. *10th suppl*. London: Roy. Hort. Soc.

————. 1992c. *The international Rhododendron register (1958)*. *32d suppl*. London: Roy. Hort. Soc.

————. 1993a. *The international lily register (1982)*. *11th suppl*. London: Roy. Hort. Soc.

————. 1993b. *The international Rhododendron register (1958)*. *33d suppl*. London: Roy. Hort. Soc.

————. 1994. *The international lily register (1982)*. *12th suppl*. London: Roy. Hort. Soc.

Royal Horticultural Society and General Dutch Bulbgrowers' Society. 1948. *A classified list of tulip names*. London: Royal Horticultural Society and General Dutch Bulbgrowers' Society.

————. 1952. *A classified list of tulip names*. London: Roy. Hort. Soc., and Haarlem, Netherlands: Gen. Dutch Bulbgrowers' Soc.

Royal National Rose Society. 1976. *Roses: a selected list of varieties*. 6th ed. St. Albans, England: Roy. Natl. Rose Soc.

Rubtsov, L. I., V. G. Zhogoleva, and N. A. Lyapunova. 1961. *Sad sireni* (Syingarii) (in Russian). Iedatel. Kiev: Akad. Nauk.

Rubtsov, L. I., N. L. Mikhailov, and V. G. Zhogoleva. 1982. Lilac species and cultivars in cultivation in U.S.S.R. *Lilacs* 11(2): 1–38.

Ruksans, J. 1981. *Krokusi* (in Latvian). Riga: Avots.

Rushforth, K. 1985. Hornbeams and hop hornbeams. *Plantsman* 7: 173–191, 205–212.

Russell, P. 1934. *The oriental flowering cherries*. U.S.D.A. Circ. 313.

Salley, H. E., and J. E. Greer. 1986. *Rhododendron hybrids: a guide to their origins* (includes selected, named forms of *Rhododendron* species). Portland, Oregon: Timber Press.

Salley, H. E., and J. E. Greer. 1992. *Rhododendron hybrids: a guide to their origins* (includes selected, named forms of *Rhododendron* species). 2d ed. Portland, Oregon: Timber Press.

Sampson, D. R. 1965. Breeding *Philadelphus* for double flower, purple center and low stature. *Euphytica* 14: 157–160.

Sand, A. W. W. 1925. *Bearded Iris: A perennial suited to all gardens.* Cornell Univ. Agric. Exp. Sta. Bull. 112.

Sanders. 1946. *Sanders' complete list of orchid hybrids containing the names and parentages of all the known hybrid orchids whether introduced or artificially raised to January 1st, 1946.* St. Albans, England: Sanders.

Sanders, C. R. 1982. Comments on *Hamamelidaceae. Plantsman* 4: 126–127.

Sanders, D. F., and Mrs W. J. Wreford. 1961. *David Sanders' one-table list of orchid hybrids containing the names and parentages of all registered orchid hybrids from the 1st January 1946 to 31st December 1960.* East Grinstead, England: David Sander's Orchids.

Sandhack, H. A. 1927. *Dahlien und Gladiolen: ihre Beschreibung, Kultur und Züchtung.* Berlin: Paul Parey.

Santamour, F. S., S.-a. He, and A. J. McArdle. 1983. Checklist of cultivated *Ginkgo. J. Arbor.* 9: 88–92.

Santamour, F. S., and A. J. McArdle. 1982a. Checklist of cultivated maples. I. *Acer rubrum* L. *J. Arbor.* 8: 110–112.

———. 1982b. Checklist of cultivated maples. II. *Acer saccharum* Marshall. *J. Arbor.* 8: 164–167.

———. 1982c. Checklist of cultivated maples. III. *Acer platanoides* L. *J. Arbor.* 8: 241–246.

———. 1982d. Checklist of cultivated maples. IV. *Acer saccharinum* L. *J. Arbor.* 8: 277–280.

———. 1983a. Checklist of cultivars of Callery pear *(Pyrus calleryana). J. Arbor.* 9: 114–116.

———. 1983b. Checklist of cultivars of honeylocust (*Gleditsia triacanthos* L.). *J. Arbor.* 9: 248–252.

———. 1983c. Checklist of cultivars of North American ash *(Fraxinus)* species. *J. Arbor.* 9: 271–276.

———. 1984. Cultivar checklist for *Liquidambar* and *Liriodendron. J. Arbor.* 10: 309–312.

———. 1985a. Cultivar checklist of the large–bracted dogwoods: *Cornus florida, C. kousa,* and *C. nuttallii. J. Arbor.* 11: 29–36.

———. 1985b. Checklists of cultivars of linden *(Tilia)* species. *J. Arbor.* 11: 157–164.

———. 1986. Checklist of cultivated *Platanus* (planetree). *J. Arbor.* 12: 78–83.

———. 1988. Cultivars of *Salix babylonica* and other weeping willows. *J. Arbor.* 14: 180–184.

———. 1989. Checklists of cultivars in *Betula* (birch). *J. Arbor.* 15: 170–176.

Saunders, E. 1971–1973. *Wagtails book of fuchsias. 3 vols.* Henfield, England: E. R. M. Saunders.

Savige, T. J. 1993. *The international Camellia register. 2 vols.* Int. Camellia Soc.

Schaepman, H. K. E. 1975. *Preliminary checklist of cultivated Hedera. pt. 1: juvenile varieties and cultivars of Hedera helix.* Mt Vernon, Virginia: Am. Ivy Soc.

Schilling, T. 1982. A survey of cultivated Himalayan and Sino–Himalayan *Hedychium* species. *Plantsman* 4: 129–149.

Schalk, P. H. 1985. *Sophora japonica*. *Dendroflora* 22: 69–72.

Schlauer, J. 1986. Nomenclatural synopsis of carnivorous phanerogamous plants: A world carnivorous plant list. *Carnivorous Pl. Newslett.* 15: 59–116.

———. 1987. Corrections to the world carnivorous plant list. *Carnivorous Pl. Newslett.* 16: 99–101, 103.

Schmalscheidt, W. 1980. *Rhododendron-züchtung in deutschland.* Oldenburg, Germany: W. Schmalscheidt.

———. 1984. *Potentilla* und *Pyracantha*-Ergebnisse eines Sortenvergleiches. *Deutsche Baumschule* 36(5): 202–205.

Schmid, W. G. 1991. *The genus Hosta=giboshi zoku.* Portland, Oregon: Timber Press.

Schmidt, M. G. 1962. *Ceanothus* round up. *J. Calif. Hort. Soc.* 23(2): 64–68.

Schneider, C. 1923. Notes on hybrid *Berberis* and some other garden forms. *J. Arnold Arbor.* 4: 193–232.

———. 1930. Die Gattung *Diervilla* und *Weigela*. *Mitt. Deutsch. Dendrol. Ges.* 42: 13–23.

———. 1934. Mock oranges. *New Fl. & Silva* 6: 113–117.

Schneider, F. 1965a. *Alnus, Carpinus* en *Ostrya*. *Dendroflora* 2: 7–9.

———. 1965b. *Hypericum*. *Dendroflora* 2: 18–22.

———. 1965c. Blauw bloeiende dwergrhododendrons. *Dendroflora* 2: 23–27.

———. 1966a. *Hypericum* (II). *Dendroflora* 3: 37–39.

———. 1966b. *Rhododendron williamsianum* hybriden. *Dendroflora* 3: 47–59.

———. 1967. *Potentilla fruticosa*. *Dendroflora* 4: 42–50, ill. opp. p. 65.

———. 1968. *Astilbe*. *Dendroflora* 5: 3–10.

———. 1971. *Lonicera nitida* en *Lonicera pileata*. *Dendroflora* 8: 42–45.

———. 1986a. Statutory and nonstatutory registration authorities. *Acta Hort.* 182: 219–224.

———. 1986b. The concept of distinctness in plant breeders' rights. Pp. 393–397 in B. T. Styles, ed., *Infraspecific classification of wild and cultivated plants.* Oxford: Clarendon Press.

———. and H. J. van de Laar. 1970. *Escallonia*. *Dendroflora* 7: 35–41.

Schwerin, F. Graf von. 1919. *Acht Beiträge zur Gattung Acer.* Wendisch–Wilmersdorf bei Thyrow (Kreis Teltow), Germany: Fritz Graf von Schwerin.

Scott, E.L., and A. H. Scott. 1950. *Chrysanthemums for pleasure.* 2d ed. Bogota, New Jersey: Scotts.

Sharma, J. N., and H. N. Metcalf. 1968. A horticultural classification of contemporary zinnias. *Proc. Montana Acad. Sci.* 28: 28–38.

Sharp, M. L., ed. 1957. *Camellias illustrated. rev. ed.* Portland, Oregon: Camellia Soc.

Sieber, J. 1990a. *Stauden–Neuheiten, die in die letzen 5 Jahren angemeldet wurden.* Freising, Federal Republic of Germany: J. Sieber.

———. 1990b. *Stauden–Neuheiten, die 1990 angemeldet und registriert wurden (Stand Juli 1990).* Freising, Federal Republic of Germany: J. Sieber.

Simon, L., and P. Cochet. 1906. *Nomenclature de tous les noms de roses.* Paris: Librairie Horticole.

Singer, M. 1885. *Dictionnaire des roses.* Berlin: Paul Parey.

Singh, B. 1986. New *Bougainvillea* cultivars registered. *Bougainvillea Newsl.* 6(2): 17–18.

Sitch, P. 1975. *Carnations for garden and greenhouse.* London: John G. Gifford.

Smee, S. 1984. Growing and breeding nerines. *The Garden* (London) 109: 408–413.

Smith, F. C. 1990. *A plantsman's guide to carnations and pinks.* London: Ward Lock.

Smith, M. N. 1979. *Ceanothus* of California: A gardener's guide. *Pacific Hort.* 40(2): 37–45, (3): 36–43.
Smithers, P. 1993. *Nerine sarniensis. The Garden* (London) 118: 190–193.
Society for Japanese Irises. 1988. *The 1988 cumulative checklist of Japanese irises.* Soc. Japanese Irises.
Sorvig, K. 1983. The genus *Nigella. Plantsman* 4: 229–235.
South African Aloe Breeders Association. 1971. *Check list of Aloe cultivars.* Karen Park: South African Aloe Breeders Assoc.
Spingarn, J. E. 1935. The large–flowered *Clematis* hybrids. *Natl. Hort. Mag.* 14: 64–91.
Spongberg, S. A. 1988. Cultivar registration at the Arnold Arboretum. *HortScience* 23: 456–458.
——. 1989. Cultivar name registration at the Arnold Arboretum 1987 and 1988. *HortScience* 24: 433–434.
——. 1990. Cultivar registration at the Arnold Arboretum. *HortScience* 25: 617.
——. 1991. Cultivar registration at the Arnold Arboretum 1990. *HortScience* 26: 476.
——. 1992. Cultivar registration at the Arnold Arboretum 1991. *HortScience* 27: 303.
——. 1993. Cultivar registration at the Arnold Arboretum 1992. *HortScience* 28: 280.
——. 1994. Cultivar registration at the Arnold Arboretum 1993. *HortScience* 29: 971.
Sprenger, C. 1911. Neue Mitteilungen über *Wistaria chinensis* DC. *Mitt. Deutsch. Dendrol. Ges.* 20: 237–240.
Starcs, K. 1928. Ubersicht über die Arten der Gattung *Syringa* L. *Mitt. Deutsch. Dendrol. Ges.* 40: 31–49.
Stock, K. L. 1984. *Rose books.* Milton Keynes, England: K. L. Stock.
Stout, A. B. 1916. The development of the horticultural varieties of *Coleus. J. New York Bot. Gard.* 17: 209–218.
——. 1934. *Daylilies.* New York: Macmillan.
——. 1986. *Daylilies.* Millwood, New York. Sagapress.
Stover, H. 1983. *The Sansevieria book.* Tustin, California: Endangered Species Press.
Straley, G. B. 1984. The Kaffir lily. *Pacific Hort.* 45(4): 25–27.
Swartley, J. C. 1984. *The cultivated hemlocks.* Portland, Oregon: Timber Press.
Swinbourne, R. F. G. 1979. *Sansevieria in cultivation in Australia.* Adelaide Bot. Gard. Handb. 2.
Swindells, P. 1983. *Waterlilies.* London: Croom Helm.
——. 1989a. *A plantsman's guide to primulas.* London: Ward Lock.
——. 1989b. *A preliminary checklist of cultivated Nymphaea.* Int. Water Lily Soc.
Symons–Jeune, B. H. B. 1953. *Phlox: a flower monograph.* London: Collins.
Taylor, J. 1985a. *Kniphofia*—A survey. *Plantsman* 7: 129–160.
——. 1985b. The national *Kniphofia* collection. *Hardy Pl. Soc. Bull.* 7(1): 8–13.
Thomas, G. S. 1990. *Perennial garden plants or the modern florilegium.* Portland, Oregon: Sagapress.
Thompson, M. L. 1976–1978. *The Thompson Begonia guide. 3 vols. + suppl. 2d ed.* Southampton, New York: Edward J. Thompson.
——. 1984. *Begonias: 1984 update.* Southampton, New York: E. J. Thompson.
Thompson, M. L., and E. J. Thompson. 1980. *The Thompson Begonia guide. 2d ed. suppl. 1. exhibition manual.* Southampton, NY: Edward J. Thompson.
—— 1981. *Begonias: the complete reference.* New York: Times Books.

—— 1982. *The Thompson Begonia guide. 2d ed, rev. with additional sheets. suppl.*
    1. exhibition manual. Southampton, NY: Edward J. Thompson.
Thorne, T. 1959. *Fuchsias for all purposes.* London: W. H. & L. Collingridge.
Thornton, E. H., and S. H. Thornton. 1985. *The exotic Plumeria (frangipani).*
    Houston, Texas: Plumeria Specialities.
Throckmorton, T. D. [Not dated] *The daffodil data bank of the A.D.S.* Am. Daffodil
    Soc.
Tompsett, A. A. 1982. *Daffodil varieties.* U.K. Minist. Agric. Fish. Food Ref. Book
    350.
Toubøl, U. 1981. Clonal variation in *Anemone nemorosa. Plantsman* 3: 167–174.
Touw, A. 1963. Een voorlopig overzicht van de Nederlandse iepen. *Jaarb. Ned.*
    Dendrol. Ver. 22: 57–62.
Townsend, K. J. 1984. *Achimenes*—the hot water plants. *Plantsman* 5: 193–205.
Traub, H. P. 1961. The genus ×*Crinodonna* 1921–1960. Catalog of ×*Crinodonna*
    cultivars. *Pl. Life* 17: 65–74.
Traub, H. P., W. R. Ballard, L. F. Morton, W. D. Morton, and E. T. Authement.
    1964. Catalog of hybrid *Amaryllis* cultivars 1799 to Dec. 31, 1963. *Pl. Life*
    20 (Suppl.): 1–21.
Traub, H. P., and L. S. Hannibal. 1960. Catalog of *Brunsvigia* cultivars 1837–1960.
    Pl. Life 16: 39–62.
Trehane, P. 1989. *Index hortensis. Vol. 1: Perennials.* Wimborne, England:
    Quarterjack.
Treseder, N. G. 1978. *Magnolias.* London: Faber & Faber.
Tretter, A. 1976. *African violet master variety list 1966–1976. registered varieties*
    1948–1976. Knoxville, Tennessee: African Violet Soc. Am.
Tripp, K. E. 1993. Sugi. *Am. Nurseryman* 178(7): 26–39.
Tucker, A. O., and K. J. W. Hensen. 1985. The cultivars of lavender and lavandin.
    Baileya 22: 168–177.
Tucker, A. O., and M. J. Maciarello. 1986. The essential oils of some rosemary
    cultivars. *Flavour & Fragrance J.* 1: 137–142.
Tucker, A. O., and E. D. Rollins. 1989. The species, hybrids, and cultivars of
    Origanum (Lamiaceae) cultivated in the United States. *Baileya* 23: 14–27.
Turner, R. 1983. A review of spurges for the garden. *Plantsman* 5: 129–156.
Tuyama, T. 1968. *Camellias of Japan. 2 vols.* Osaka, Japan: Takeda Sci. Found.
Underhill, T. L. 1990. *Heaths and heathers: the grower's encyclopedia.* London:
    David & Charles.
Unwin, C. W. J. 1926. *Sweet peas: their history, development and culture.* New York:
    D. Appleton.
Vail, R. 1984. *Platycerium hobbyist's handbook.* Dona Ana, New Mexico: Desert
    Biol. Publ.
Valavanis, W. N. 1976. *The Japanese five needle pine: Nature, gardens, bonsai,*
    taxonomy. Encyclopedia of classical bonsai art. Vol. 2. Atlanta, Georgia,
    Symmes Systems.
Vanderplank, J. 1991. *Passion flowers and passion fruit.* Cambridge, Massachussets:
    MIT Press.
Van Eseltine, G. P. 1933. Notes on the species of apples. II. The Japanese flowering
    crabapples of the sieboldii group and their hybrids. *New York Agric. Exp.*
    Sta. Techn. Bull. 214: 1–21.
——. 1934. Ornamental apples and crabapples. *New York Agric. Exp. Sta. Circ.*
    139: 1–13.
Van Rensselaer, M., and H. E. McMinn. 1942. *Ceanothus.* Santa Barbara, California:
    Santa Barbara Bot. Garden

Vertrees, J. D. 1978. *Japanese maples*. Portland, Oregon: Timber Press.
———. 1979. Notes on variants of *Acer circinatum* Pursh. *Int. Dendrol. Soc. Yearb.* 1978: 82–84.
———. 1987. *Japanese maples*. 2nd ed. Portland, Oregon: Timber Press.
Vogel, P. de. 1969. *Pernettya:* Keuringen 1968. *Dendroflora* 6: 75–76.
Vogts, M. 1982. *South Africa's Proteaceae: know them and grow them*. Melbourne: Proteaflora Enterprises.
Vrugtman, F. 1972. Bibliography of cultivar name registration. *Chron. Hort.* 12(3): 47–50.
———. 1973. Bibliography of cultivar name registration, Add. 1. *Chron. Hort.* 13(3): 54.
———. 1977. Bibliography of cultivar name registration. Add. 2. *Chron. Hort.* 17(2): 29–30.
———. 1981. Bibliography of cultivar name registration. Add. 3. *Chron. Hort.* 21(2/3): 29–31.
———. 1984. Directory of International Registration Authorities for cultivar names. *Chron. Hort.* 24(1): 4–6.
———. 1985. Directory of International Registration Authorities for cultivar names. Addenda & corrigenda. *Chron. Hort.* 25(3): 42.
———. 1986. Directory of International Registration Authorities for cultivar names: Further addenda & corrigenda. Hamilton, Canada: Roy. Bot. Garden
———. 1988. Lilac registration 1986–87. *HortScience* 23: 458.
———. 1989a. Lilac registration 1988. *HortScience* 24: 435–436.
———. 1989b. Corrigenda. Lilac registration 1986–1987. *HortScience* 24: 436.
———. 1989c. Directory of International Registration Authorities for cultivar names: Further addenda & corrigenda. Hamilton, Canada: Roy. Bot. Garden
———. 1990a. Lilac registration 1989. *HortScience* 25: 618.
———. 1990b. Directory of International Registration Authorities for cultivar names. *HortScience* 25: 618–621.
———. 1990c. Addenda & corrigenda to the tentative international register of cultivar names in the genus *Syringa* L. (1976). Hamilton, Canada: Roy. Bot. Garden
———. 1990d. Directory of International Registration Authorities for cultivar names. *Chron. Hort.* 30: 60–62.
———. 1991. Lilac registration 1990. *HortScience* 26: 476–477.
———. 1994a. International registration of cultivar names for unassigned woody genera 1993. *HortScience* 29: 970–971.
———. 1994b. Lilac registration 1993. *HortScience* 29: 972.
Wagenknecht, B. L. 1961a. Registration lists of cultivar names in *Gleditsia* L. *Arnoldia* 21: 31–34.
———. 1961b. Registration lists of cultivar names in the genus *Pieris* D. Don. *Arnoldia* 21: 47–50.
———. 1965. Registration lists of cultivar names in *Buxus* L. *Boxwood Bull.* 4: 35–41.
———. 1967. Addenda to the registration lists of cultivar names in *Buxus* L. *Boxwood Bull.* 7: 1.
———. 1971. Addenda to the registration lists of cultivar names in *Buxus* L. *Boxwood Bull.* 11: 1.
———. 1972. Addenda to the registration lists of cultivar names in *Buxus* L. *Boxwood Bull.* 11: 45.
Walsweer, A. 1988. *Geraniumgids*. Heerde, Netherlands: A. Walsweer.
Wandell, W. N. 1989. *Handbook of landscape tree cultivars*. Gladstone, Illinois: East Prairie Publ.

Warburg, O., and E. F. Warburg. 1930. A preliminary study of the genus *Cistus*. *J. Roy. Hort. Soc.* 55: 1–52.

———. 1931. *Cistus* hybrids. *J. Roy. Hort. Soc.* 56: 217–224.

Warburton, B. 1986. *Check list. Society for Siberian irises. section II. varietal listing.* Soc. Siberian Irises.

Weaver, R. E. 1976a. The witch hazel family (*Hamamelidaceae*). *Arnoldia* 36: 69–109.

———. 1976b. Selected maples for shade and ornamental planting. *Arnoldia* 36: 146–176.

Webb, D. A., and R. J. Gornall. 1989. *A manual of saxifrages.* Portland, Oregon: Timber Press.

Weber, C. 1963. Cultivars in the genus *Chaenomeles*. *Arnoldia* 23: 17–75.

Webber, S., ed. 1988. *Daylily encyclopedia.* Damascus, Maryland: Webber Garden.

Wehrhahn, H. R. 1931. *Die Gartenstauden.* 2 vols. Berlin: Paul Parey.

Welch, H. J. 1979. *Manual of dwarf conifers.* Little Compton, Rhode Island: Theophrastus.

———. 1991. *The conifer manual.* Vol. 1. Dordrecht, Germany: Kluwer Acad. Publ.

Welch, H. J., and G. Haddow. 1993. *The world checklist of conifers.* Combe Martin, U.K.: World Conifer Datapool.

Wellensiek, S. J. 1961. The breeding of diploid cultivars of *Cyclamen persicum. Euphytica* 10: 259–268.

Wellensiek, S. J., J. Doorenbos, J. van Bragt, and R. A. H. Legro. 1961. *Cyclamen: a descriptive list of cultivars.* Wageningen: Laboratorium voor Truinbouwplantenteelt, Landbouwhogeschool.

Wemyss-Cooke, J. *Primulas old and new: auriculas, primulas, primroses, polyanthus.* London: David & Charles.

Werken, H. van de. 1988. Mutant offspring: Five new forsythia cultivars are the progeny of one irradiated *Forsythia* ×*intermedia* 'Lynwood Gold.' *Am. Nurseryman* 167(1): 127–132.

Wilfret, G. J. 1984. Caladiums to know and grow. *Foliage Dig.* 7(7): 1–3.

Wilson, E. H. 1916. *The cherries of Japan.* Publ. Arnold Arbor. No. 7.

———. 1923. The hortensias. *Hydrangea macrophylla* DC. and *Hydrangea serrata* DC. *J. Arnold Arbor.* 4: 233–246.

Winters, H. F. 1973. New impatiens from New Guinea. *Am. Hort.* 52(3): 16–22.

Wister, J. C. 1927. A lilac checklist. *Natl. Hort. Mag.* 6: 1–16.

———. 1942. *Lilacs for America.* Swarthmore, Pennsylvania: Arthur Hoyt Scott Hort. Found., Swarthmore College.

———. 1943. *Lilacs for America. rev. ed.* Swarthmore, Pennsylvania: Arthur Hoyt Scott Hort. Found., Swarthmore College.

———. 1953a. *Lilacs for America.* Swarthmore, Pennsylvania: Arthur Hoyt Scott Hort. Found., Swarthmore College.

———. 1953b. Preliminary holly check list. *Bull. Holly Soc. Am.* 6: 1–56.

———. 1954. Swarthmore plant notes: *A record of all plants grown by the Arthur Hoyt Scott Horticultural Foundation, Swarthmore College, Swarthmore, Delaware County, Pennsylvania, in its first twenty-five years 1930–1954. 3 vols.* Swarthmore, Pennsylvania: Arthur Hoyt Scott Hort. Found., Swarthmore College.

———. 1963. Supplementary registration list of cultivar names in *Syringa* L. registered 1963. *Arnoldia* 23: 77–83.

———, ed. 1962. *The peonies.* Washington, DC: Am. Hort. Soc.

Wolf, E. 1923. *Sambucus racemosa* Linné. *Mitt. Deutsch. Dendrol. Ges.* 33: 24–31.

Woodroof, W. E., and W. W. Donnan, ed. 1990. *Camellia nomenclature. 20th rev. ed.* Arcadia, California: Southern California Camellia Soc.
Wright, A. H. 1927. *The dwarf bearded Iris: I. A preliminary working list.* Ithaca: Cornell Univ. Agric. Exp. Sta.
Wright, D. 1980. *Philadelphus. Plantsman* 2: 104–116.
——. 1981. *Sorbus:* A gardener's evaluation. *Plantsman* 3: 65–98.
——. 1982. *Hamamelidaceae:* A survey of the genera in cultivation. *Plantsman* 4: 29–53.
——. 1983a. *Eucryphia, Hoheria* and *Plagianthus. Plantsman* 5: 167–185.
——. 1983b. Climbing honeysuckles. *Plantsman* 4: 236–252.
——. 1985. *Aesculus* in the garden. *Plantsman* 6: 228–247.
——. 1986. Notes on *Carpinus* and *Ostrya. Plantsman* 7: 212–216.
Wright, M. 1984. *The complete handbook of garden plants.* New York: Facts on File.
Wyman, D. 1943. *Crab apples for America.* Jamaica Plain, Massachusetts: Am. Assoc. Bot. Gard. and Arnold Arbor.
——. 1955. *Crab apples for America.* Am. Assoc. Bot. Gard.
——. 1958. The shrub altheas. *Arnoldia* 18: 45–51.
——. 1960. *Ilex crenata* and its varieties. *Arnoldia* 20: 41–46.
——. 1961a. Registration lists of cultivated names of forsythias. *Arnoldia* 21: 39–42.
——. 1961b. Forsythias. *Am. Hort. Mag.* 40: 190–197.
——. 1961c. Many forms of four arborvitae species grown in the trade. *Am. Nurseryman* 113(7): 10–11, 67–73, 76–79.
——. 1962a. The majestic beeches. *Arnoldia* 22: 1–7.
——. 1962b. Barberries. *Arnoldia* 22: 9–16.
——. 1962c. The birches. *Arnoldia* 22: 17–23.
——. 1962d. Hawthorns. *Arnoldia* 22: 25–32.
——. 1962e. The lindens. *Arnoldia* 22: 69–76.
——. 1962f. The oaks. *Arnoldia* 22: 77–87.
——. 1963a. International plant registration. *Arnoldia* 23: 85–92.
——. 1963b. New plants registered. *Arnoldia* 23: 111–118.
——. 1964. Registration list of cultivar names of *Fagus* L. *Arnoldia* 24: 1–8.
——. 1965. The mock–oranges. *Arnoldia* 25: 29–36.
——. 1966. More plant registrations. *Arnoldia* 26: 13–16.
——. 1967. More plant registrations. *Arnoldia* 27: 61–66.
——. 1968. *Potentilla fruticosa* varieties in the Arnold Arboretum. *Arnoldia* 28: 125–131.
——. 1969a. Plant registrations. *Arnoldia* 29: 1–8.
——. 1969b. The mountain–ashes. *Arnoldia* 29: 61–68.
Yeo, P. F. 1964. *Lonicera pileata* and *L. nitida* in cultivation. *Baileya* 12: 56–66.
——. 1971a. Further observations on *Bergenia* in cultivation. *Kew Bull.* 26: 47–56.
——. 1971b. Cultivars of *Bergenia* (*Saxifragaceae*) in the British Isles. *Baileya* 18: 96–112.
——. 1975. The hybrid origin of some cultivated snowdrops (*Galanthus – Amaryllidaceae*). *Baileya* 19: 157–162.
——. 1985. *Hardy geraniums.* London: Croom Helm.
Yinger, B. R., and G. R. Hahn. 1985. Cultivars of Japanese plants at Brookside—II. *Arnoldia* 45(2): 7–18.
Yokoi, M., and Y. Hirose. 1978. *Variegated plants.* Japan: Seibundo Shinkosa Publ.

APPENDIX XII

GLOSSARY

This glossary provides succinct meanings of terms used in this *Code,* and also provides reference for other terms which are commonly used in discussions on the nomenclature of cultivated plants.

For a fuller glossary of terms used in nomenclature generally, the user is referred to Hawksworth, D. L. (ed.), *A draft glossary of terms used in bio-nomenclature,* IUBS Monograph No 9, published by the International Union of Biological Sciences, Paris, 1994.

abbreviation
    a shortening of a word or words using selected letters.
abstracting journal
    one which publishes summaries or abstracts of articles published elsewhere.
accepted name
    the name to be used in accordance with the rules of a *Code* for a taxon in any given taxonomic position.
acronym
    an abbreviation consisting of the initial letters of several words.
addition sign (+)
    the indication of a graft–chimaera.
admissible epithet
    an unpublished epithet that is in a form that would permit it to be established.
agamospermy
    a plant which is able to produce viable seeds without fertilization, and whose progeny is, therefore, usually genetically uniform.
agriculture
    the extensive cultivation of non–woody food, fodder and industrial crops.
allopolyploid
    a polyploid with chromosomes sets derived from different species.
alternative terminology
    differing terms used for the same concept.
apomict
    a plant habitually reproducing by asexual means, e. g. vegetatively, or by producing viable seed (agamospermy) or spores (apospory) without fertilization, and whose progeny is, therefore, usually genetically uniform.
apospory
    a plant which is able to produce viable spores without fertilization, and whose progeny is, therefore, usually genetically uniform.

appropriate epithet
> one of which the etymology relates to some aspect of the taxon, for example its characters, provenance, or history.

arabization
> the writing of sounds, signs or letters of a non–Arabic writing system in Arabic letters.

arbitrary epithet
> one formed without regard to etymology or other formal derivation.

archetype
> the ideal or theoretical manifestation of a taxon.

article
> a rule in a *Code* which is binding.

artificial classification
> one that is proposed solely for ease of use and which does not pretend to demonstrate a natural order.

ascription
> the direct association of the name of a person or persons with publication of a new name (or epithet) or with the description or diagnosis of a taxon.

asexual propagation (vegetative propagation)
> the propagation of organisms without fertilization such as by means of agamospermy, cuttings, division, meristem culture and micropropagation.

assignee
> one who is appointed to act for an originator.

authentic specimen
> one that has been verified by an author of a name as being representative of the taxon to which the name is assigned.

author
> the person to whom a name, epithet, or publication is attributed.

author abbreviation
> an abbreviation of the name of an author used in an author citation to save space.

author citation
> the name(s) of the author(s) responsible for introducing a name or epithet to the name or epithet itself.

autograph
> a text written by hand and not reproduced by any machine.

autopolyploid
> a polyploid with chromosome sets derived from a single species or individual.

avowed substitute epithet
> an epithet explicitly proposed as a substitute for an unacceptable epithet.

back–cross
> the cross of a hybrid with one of its own parents.

basionym
> the name (of a genus) or epithet of another taxon which was originally published in a different taxonomic position to that currently used.

bigeneric hybrid
> one resulting from the sexual union of individuals assigned to different genera.

binary name – See: **binomen**
binomen
the scientific name of a species consisting of two words. The first word is the name of the genus to which the species belongs and the second is the epithet given to that species to distinguish it from others in the same genus.
binomial – See: **binomen**
bonsai
the oriental art of dwarfing trees and other plants.
Botanical Code – See: **International Code of Botanical Nomenclature**
botanical Latin – See: **Latin**
botanical name – See: **scientific name**
brackets
a pair of marks such as round brackets "()" (in North America often called parentheses), square brackets "[]" or curly brackets "{}" used as parentheses for enclosing words or numbers so as to separate them from their context.
breeder
an originator of seed-raised cultivated plants.
budwood
shoots of scion material from which buds are obtained for budding.
category
a concept by which we can understand or classify matter: not to be confused with "taxon".
character
any one feature of the organism under consideration.
chimaera
an individual composed of two or more genetically different tissues.
circumscribe
to make a circumscription.
circumscription
a statement of the diagnostic limits of a taxon, which, in the opinion of an author, separate the individuals that belong to it from all other individuals.
classification
the systematic grouping of items: a system in which items may be grouped.
clone
an indefinite number of ramets derived from an ortet through asexual propagation and which should remain genetically identical.
Code
one of the international codes of nomenclature, generally referring to its most recent edition.
code-name
a linguistically meaningless epithet made up of an arbitrary sequence of connected letters and/or numerals.
coextension
a situation when two or more taxa contain the same members and no others.
cognomen
a nickname for a person: a shorthand name used to label material of uncertain identity.

collective epithet
    the epithet in Latin or a modern language which designates a nothotaxon below the rank of nothogenus.
collective name
    the single designation covering the progeny of a particular hybrid combination.
colloquial name
    a name that is used locally but not widely enough to be recognized in the general dictionaries of the language concerned. See also: **common name** and **vernacular name**
colour mixture
    seed from field–grown, open–pollinated lines, which may or may not be mixed to a particular formula.
commercial synonym – See: **trade designation**
common name
    one widely used in any language for a taxon and which is generally to be found in non–technical dictionaries of that language. See also: **colloquial name** and **vernacular name**
competing epithet
    epithets taken into consideration in determination of the accepted epithet for a taxon.
complex
    a group of individuals including more than a single taxon, but within which the circumscription of those taxa is not satisfactorily resolved.
compound word
    one that is formed by the union of two or more basic components from other words, usually excluding prefixes and suffixes.
condensed formula
    the designation of a nothogenus or graft–chimaeral genus constructed from parts of the names of the parent genera involved.
congeneric
    belonging to the same genus.
connecting vowel
    one that joins different word elements.
consonant
    in a particular language, the sound in speech, produced by a complete momentary stoppage or by partial stoppage or constriction of the air–stream in some part of the mouth–cavity or the lips as it passes from the lungs: a letter or symbol expressing such a sound.
conspecific
    belonging to the same species.
contraction
    an abbreviation that shortens a word by removing one or more of its middle letters.
convar
    or convarietas, a category of cultivated plants which is placed within the ranks of a nomenclatural hierarchy.
correct name – See: **accepted name**

covariant
    a plant whose characteristics are maintained solely by regular practices of
    cultivation such as by pruning or training.
cross
    to cause to interbreed (verb): the act of hybridization (subst.).
cryptic variety
    a cultivated plant with attributes that cannot be readily taxonomically
    separated from a cultivar.
cultigen
    a species believed not to have originated in the wild.
cultivar
    a taxon of cultivated plants that is clearly distinct, uniform and stable in its
    characteristics and which, when propagated by appropriate means, retains
    those characteristics.
cultivar class – See: **denomination class**
cultivar epithet
    the defining part of a name that denotes a cultivar.
cultivar-group
    a taxon of cultivated plants that denotes an assemblage of similar named
    cultivars.
cultivariant
    a topophysic plant.
cultivated plant
    one whose origin or selection is primarily due to the intentional activities of
    mankind. Such a plant may arise either by deliberate or, in cultivation,
    accidental hybridization, or by selection from existing cultivated stock, or
    may be a selection from minor variants within a wild population and
    maintained as a recognisable entity solely by deliberate and continuous
    propagation.
Cultivated Plant Code – See: **International Code of Nomenclature for
    Cultivated Plants**
culton (plural: culta)
    a systematic group of cultivated plants which is based upon one or more
    user criteria: a word parallel to "taxon" but used solely for a taxonomic
    group of cultivated plants.
cyclophysic plants
    two or more plants with the same genotype but differing in their phenotype
    due to being originally asexually propagated from different phases in the
    plant's life-cycle.
cyrillization
    the writing of sounds, signs or letters of a non-Cyrillic writing system in
    Cyrillic letters.
cytological characters
    the characteristics within the cellular structure of an organism.
date
    of a cultivar or cultivar-group epithet, that of its establishment.

date of publication
> the actual time (day, month and year, or month and year, or at least the year), on which printed matter became available to the general public or to botanical libraries.

denomination – See: **varietal denomination**

denomination class
> the taxon in which cultivar epithets may not be duplicated except in special circumstances.

derived variety
> a cultivar that is derived from an earlier open–pollinated cultivar and which no longer meets the criteria of a DUS test.

description
> a statement of the characters of a particular taxon: an expanded diagnosis.

descriptor
> a word or phrase used to denote a varietal component isolated from a series.

determine
> to perform an identification.

determination
> an identification.

determination slip
> a piece of paper attached to a specimen on which is written the determination, along with details of who made the determination, the date of determination, and other notes that may be made at the time of determination.

diacritic
> a mark, such as an accent or diaeresis connected with a letter, used to indicate pronunciation.

diaeresis
> a diacritic (¨) placed over the second of two consecutive vowels to indicate that they are to be pronounced separately.

diagnosis (plural: diagnoses)
> a statement which, in the opinion of its author, distinguishes a taxon from others.

diagnostic characters
> features that permit one to assign a plant to a particular taxon or group of taxa.

differential character
> a distinguishing or diagnostic character.

disagreeable epithet
> one whose use in a certain culture might cause offence.

diverged variety
> a cultivar that is derived from an earlier self–pollinated cultivar and which no longer meets the criteria of a DUS test.

double–cross
> the repeatable crossing of two F_1 hybrids.

duplicate
> of a specimen, part of a single gathering from a plant made by a collector at one time.

DUS test
  the criteria of Distinctness, Uniformity and Stability by which a new cultivar is examined for statutory purposes such as for the granting of Plant Breeders' Rights.
effective publication – See: **publication**
elision
  the omission of a letter or word.
ephemeral
  of publications, ones not intended for continued reference, such as newspapers and advertisements, which ordinarily become obsolete soon after they are issued.
epithet
  the final word or combination of words in a name that denotes an individual taxon: in a *Code*, one that is established under the rules of that *Code* whether or not it is an accepted or acceptable epithet. See also: **cultivar epithet**
epitype
  under the *ICBN*, the specimen or illustration selected to serve as interpretative type when the holotype, lectotype or previously designated neotype, or all original material associated with an established name, is demonstrably ambiguous and cannot be critically identified for purposes of the precise application of the name of a taxon.
eponym
  a name based on that of a person or persons.
equivalent epithet
  one that is an exact translation or transliteration of an existing epithet.
established
  of a name or epithet, one that fulfills the requirements of establishment.
establishment
  a prime principle of nomenclature in a *Code* whereby, on publication, certain criteria must have been fulfilled before considerations of acceptability are entertained.
etymology
  the original linguistic derivation of a word.
exemplar
  a representative sample of a taxon.
extant
  of a taxon, having living representatives.
extinct
  of a taxon, having no living representatives.
FAO
  the acronym for the Food and Agriculture Organization of the United Nations.
F$_1$ hybrid (single cross)
  a plant breeding term for the result of a repeatable single cross between two pure-bred lines.
F$_2$ hybrid
  a plant breeding term for the result of self-pollination within a population of F$_1$ hybrids.

fancy epithet
a cultivar epithet written in a modern language and not using Latin.
family (*familia*)
the principal category in the nomenclatural hierarchy between order (*ordo*)
and genus (*genus*).
forestry
the extensive cultivation of woody plants in woodlands.
form (*forma*)
the secondary category in the nomenclatural hierarchy below variety (*varietas*).
formula mixture
a seed mixture made from known components to a pre–determined formula
for any defined market.
four-way cross
the repeatable crossing of four F_1 hybrids.
gender
a grammatical characteristic of nouns in some languages that may determine
the way in which they, and adjectives, articles and pronouns applied to
them, are inflected; in Latin, Greek and many modern languages there are
three genders; masculine, feminine and neuter.
gene bank
A place where genetic variants are physically stored.
gene pool
the range of genetic variation found in a population.
generic
of taxa, pertaining to a genus.
generic designation
a name referring to any multiplicity of objects having one or more properties in common which distinguishes it from other groups of objects.
generic name
the name of a genus: a legal term for the commonly–used name of goods.
genetically modified plant (GMP)
a plant with new characters following the implantation of alien genetic
material.
genotype
the genetic make–up of an individual.
genus (*genus*)
the principal category in the nomenclatural hierarchy between family
(*familia*) and species (*species*).
GMP – See: **Genetically Modified Plant**
graft–chimaera
a taxon of cultivated plants whose members consist of tissue from two or
more different taxa in intimate association, effected by grafting.
graft hybrid – See: **graft–chimaera**
grafting
the (usually deliberate) fusion of tissue from two or more different plants.

grex
  a collective term applied to the progeny of an artificial cross from known parents; each and every crossing of any two parent plants belonging to different taxa that bear the same pair of specific, intraspecific, infraspecific, grex or cultivar epithets.

group
  to assemble a number of items together (verb): an informal assemblage of taxa (subst.)

Group – See: **cultivar–group**

hand–pollination
  the controlled act of pollination that should exclude the possibility of open–pollination.

handwriting
  the immediate product of a person's writing.

Hepburn
  the international standard system for Japanese latinization.

herbarium (plural: herbaria)
  a collection of (usually dried) botanical specimens: the housing for such specimens.

herbarium specimen
  a (usually dried) botanical specimen kept in a herbarium.

heterogen
  a variable group of individuals arising as hybrids, sports or other mutations.

heterogenous taxon
  one which, as circumscribed, includes two or more elements which are considered to belong to one or more other taxa.

heterotypic synonym
  under the Rules of the *ICNB*, a name which, in the opinion of an author, has a name–bearing type belonging to the same taxon as another for which a name has previously been established.

hierarchy
  the categories of taxa arranged in order according to their rank.

holograph
  a handwritten document.

holotype
  under the *ICBN*, the one specimen or illustration used by the author, or designated by the author, as the nomenclatural type of a name of a species or infraspecific taxon. As long as the holotype is extant, it fixes the application of the name concerned.

homonym
  one of two or more names or epithets spelled, or deemed to be spelled, exactly like another name or epithet, but which is used for different taxa of the same rank. See also: **parahomonym**

homotypic synonym
  under the Rules of the *ICBN*, a name that shares the same name–bearing type as an earlier name.

horticulture
  the cultivation of ornamental, fruit and vegetable plants.

hybrid
the result of a cross between differing plants or taxa. See also: **nothotaxon**
hybrid binomen
a collective name covering all the progeny of a particular nothospecies.
hybrid formula
the names of the parent taxa of a nothotaxon linked with a multiplication sign.
hyphen
a mark (–) used to link together two or more words so as to form a single word.
hyponym
a name that cannot be recognised because of the vagueness or inadequacy of the original description or diagnosis.
IAPT – See: **International Association for Plant Taxonomy**
ICBN – See: **International Code of Botanical Nomenclature**
ICNCP – See: **International Code of Nomenclature for Cultivated Plants**
ICNCP – See: **International Commission for the Nomenclature of Cultivated Plants**
identification
the matching of a plant or specimen with a taxon.
ideograph
the distinct character for each mental concept used in ideographic writing systems such as Chinese.
ideotype
the conceptual ideal or archetype of a series.
illegitimate
an obsolete term formerly applied to names or epithets not in accordance with certain of the rules of nomenclature; still used in the *ICBN* for names that are later homonyms or were nomenclaturally superfluous when published.
inappropriate name
a name with unsuitable etymology but which may not be rejected on that account.
inbreeding
the production of offspring between closely related parents leading to a high degree of similarity; self–fertilization is the most intense form of inbreeding.
indelible autograph
handwritten material reproduced by some mechanical or graphic process such as lithography, offset, or metallic etching.
indigen
a species believed to originate in a wild habitat.
infrageneric
pertaining to any taxon below the rank of genus.
infraspecific
pertaining to any taxon below the rank of species.
intergeneric – See: **nothogeneric**
interspecific – See: **nothospecific**

International Association for Plant Taxonomy
the organization promoted by the **International Union of Biological Sciences** (IUBS), charged with preparing the International Code of Botanical Nomenclature.

International Code of Botanical Nomenclature (Botanical Code or *ICBN*)
the book containing the international set of rules that provides for the formation and use of the scientific names of plants in Latin.

International Code of Nomenclature for Cultivated Plants (Cultivated Plant Code or *ICNCP*)
the book containing the international set of rules that provides for the formation and use of the scientific names of cultivated plants using either Latin or fancy names.

International Commission for the Nomenclature of Cultivated Plants.
an organization promoted by the International Union of Biological Sciences (IUBS) to formulate the International Code of Nomenclature for Cultivated Plants (ICNCP).

International Commission for Nomenclature and Registration
an organization promoted by the International Society for Horticultural Science (ISHS) to promote registration of the names of cultivated plants.

International Registrar
the person appointed by an International Registration Authority to carry out registration activity on its behalf.

International Registration Authority (IRA)
the organization appointed by the ISHS International Commission for Nomenclature and Registration as being responsible for registering cultivar and cultivar-group epithets within defined taxa.

International Society for Horticultural Science (ISHS)
an organization promoted by the International Union of Biological Sciences (IUBS) to promote the science of horticulture.

International Union of Biological Sciences (IUBS)
an organization promoted by UNESCO to promote all disciplines within life science.

interspecific – See: **nothospecific**

intraspecific
pertaining to the progeny of two or more individuals of the same species.

introducer
of a cultivar, the person or organization who first distributes a cultivar.

invalid
under the *ICBN*, a designation that is not established, i.e. is not strictly a "name".

ISHS – See: **International Society for Horticultural Science**

ISO
acronym for the International Organization for Standardization

isotype
under the *ICBN*, any duplicate of the holotype: it is always a specimen.

italic
the printer's type face often used to demarcate botanical names used in a formal taxonomic sense.

IUBS – See: **International Union of Biological Sciences**

kanji
the Chinese characters used as ideographs in Oriental writing.
journal
a serial publication issued at regular intervals.
juxtaposition
the placement of more than one item side by side.
Latin (botanical Latin)
the language derived from Renaissance Latin and other ancient languages which is used for the international communication of plant names and other associated descriptive information.
Latin alphabet
the characters originally derived from Latin and used for writing words in most Western cultures.
latinization
the writing of signs or letters of a non–Latin alphabet writing system in Latin letters.
lectotype
under the *ICBN*, a specimen or illustration selected as the nomenclatural type when no holotype was indicated at the time of publication, or for as long as the holotype is missing.
lectotypify
to select and publish a lectotype.
legitimate
(obsolete) the opposite of illegitimate; still applied in the *ICBN* to names that are in accordance with all the rules of nomenclature, including those on homonymy and superfluity.
ligature
the typographical union of two letters of the alphabet.
line
a plant breeding term used to describe plants resulting from repeated self–fertilization or inbreeding.
logotype (logo)
a graphical device that distinguishes goods or services from others.
lumping
to treat, as members of a single taxon, elements which have been previously considered as belonging to more than one taxon.
maintainor
one who is responsible for the continuing existence of a maintenance.
maintenance (selection)
a seed–raised cryptic variety which, although not treated as a separate cultivar, may require special designation: the designation of such by a statutory plant registration authority.
Mandarin – See: **Modern Standard Chinese**
manuscript
handwritten or typewritten material existing in only one or a few copies; writing in which each letter is placed on paper in sequence by hand, whether manually or mechanically, without any subsequent multiplication.
McCune–Reischauer
the international standard for transliteration of Korean into latin script.

mericlone
  a clone propagated by meristem culture.
microfiche
  a reproduction, much reduced, of many pages copied onto single photographic film sheets.
microfilm
  a copy of a published work or manuscript, usually much reduced, onto a continuous strip of transparent photographic material.
microform
  a term including microfiche and microfilm.
misapplied name
  a name applied to a taxon in a sense that excludes the type of the name.
modern language
  one currently in use.
Modern Standard Chinese (Mandarin)
  the standard language of China which provides the basis for the transliteration of pinyin Chinese into latin script.
morphological characters
  the physical characters of a plant.
multiline
  a plant breeding term used to describe an agricultural or horticultural cultivar made up of several closely related lines.
multiplication sign (×)
  the indication of hybridity.
mutant
  an individual produced as a result of mutation: the phenotypic expression of a mutant.
mutation
  a spontaneous or engineered change in the genotype which may alter the phenotype.
name
  in a *Code*, one that is established under the rules of the *Code* whether or not it is an accepted or acceptable name. See also: **epithet**
natural classification
  one that pretends to demonstrate a natural order.
natural hybrid
  a nothotaxon produced by chance in the wild.
neotype
  under the *ICBN*, a specimen or illustration selected to serve as nomenclatural type for as long as all of the material on which the name of the taxon was based is missing.
nomenclator
  an authoritative catalogue or other listing of names, comprising accepted names and their synonyms: one who compiles such a catalogue or list.
nom. alt.
  nominum alternativum, a permitted alternative name under the rules of the *ICBN*.
nomenclatorial
  relating to a nomenclator.

nomenclatural
    relating to nomenclature.
nomenclatural filter
    a flow diagram, or other such system, indicating the sequential steps neces-
    sary to arrive at a correct name of a taxon under a code of nomenclature.
nomenclatural hierarchy
    the consecutively subordinate categories of taxa defined by a code of
    nomenclature.
nomenclatural type (type)
    under the *ICBN*, the element on which the descriptive matter fulfilling the
    conditions of establishment for a name is based, whether as the accepted
    name or as a synonym, and which fixes the application of a name. The type
    is not necessarily the most typical or representative element of a taxon.
nomenclature
    a system of names and provisions for their formation and use.
nominant
    one who invents or coins a name for something.
non–statutory registration
    registration by any organization, agency or individual entrusted with
    registration by agreement of interested parties.
notho–
    used as a prefix, from *nothos*, Greek for hybrid.
nothogeneric
    pertaining to any nothogenus.
nothogenus
    a nothotaxon between two or more genera.
nothospecies
    a nothotaxon between two or more species.
nothospecific
    pertaining to any nothospecies.
nothotaxon
    the taxon resulting from hybridization between two or more different taxa.
off–type (rogue)
    a plant showing a non–typical phenotype.
oldest name
    the earliest established name for a taxon in any rank or category.
original material
    the plant material used by the author of a name of a taxon in the prepara-
    tion of the protologue.
original spelling
    the spelling employed when a name was established.
originator
    of a cultivar, the breeder, hybridizer, raiser or discoverer who recognizes
    the potenial attributes of a plant or group of plants.
ortet
    the original single plant from which a clone ultimately derives.
orthographical error
    an unintentional spelling error.

orthographical variant
　　an alternative spelling of a word.
orthography
　　the correct spelling.
out–cross
　　hybridization of an existing population of cultivated plants which extends its
　　gene pool and which may therefore alter the overall characteristics of that
　　population.
parahomonym
　　a word or epithet so orthographically or phonetically similar to another
　　word or epithet that they are likely to be confused.
paratype
　　under the *ICBN*, a specimen cited in the protologue that is neither the
　　holotype nor an isotype, nor one of the syntypes if two or more specimens
　　were simultaneously designated as types.
parenthesis (plural: parentheses)
　　a word, phrase or sentence inserted into a passage to which it is not gram-
　　matically essential and usually marked off by brackets, dashes or commas:
　　in the USA, round brackets (...) used for this.
parenthetical author
　　the author of a basionym shown within parenthesis in an author citation.
paronym – See: **parahomonym**
particle
　　in a word, a prefix, suffix or other minute hyphenated connecting element.
patent – See: **Plant Patent**
patronymic
　　of an epithet, one involving a patronymic prefix (e. g. Dutch "van", Irish
　　"O", Scottish "Mac", "Mc", & "M'") or suffix (e. g. Russian "-vitch",
　　Danish "-sen", English and Scandinavian "-son").
pentaploid
　　a polyploid with five sets of chromosomes.
periodical
　　a publication issued at intervals.
phase
　　a distinct stage in the growth cycle of a plant.
phenotype
　　the sum total of all the characteristics of an individual plant: the expression
　　of the genotype.
phenotypic
　　pertaining to the phenotype.
physiological characters
　　the characters related to functional mechanisms of a plant such as flowering
　　induction.
pinyin
　　the international standard for the transcription of Modern Standard Chinese
　　into latin script.
Plant Breeders' Rights (Plant Variety Rights)
　　a breeder's legal protection over the propagation of a cultivar or main-
　　tenance.

plant patent
a grant of right, available in certain countries, which provides a means of control over a new plant's propagation, use and sale for a given period.
Plant Variety Rights – See: **Plant Breeders' Rights**
pollination
the act of transferring pollen from the male part of a flower to the female part of a flower.
polybrid
a group of number of hybrids with the same parental taxa.
polyploid
having more than two sets of chromosomes.
polymorph
a taxon with many variants.
polymorphic
with many forms.
population
an assemblage of individual plants of one taxon.
precedence
a prime principle of nomenclature whereby the first name or epithet takes precedence over later names or epithets for the same taxon at a particular rank or category.
preferred name
one of two or more names considered to be more appropriate than the other(s) by an author or registration authority selecting it.
prefix
a letter or group of letters attached before the main part of a word.
pre–Linnaean
a name, epithet, or work published before the starting point of plant nomenclature, (Linnaeus's *Species Plantarum*, 1 May 1753).
preprint – See: **separate**
pre–starting point
of names or works published prior to the starting point date of the group to which the name or work belongs.
principle
in a *Code*, a fundamental precept which the Articles of a *Code* are designed to satisfy.
printed matter
text or illustrations mechanically reproduced by printing in considerable quantities and in intentionally permanent form.
printing
a process for producing identical copies by transferring an image of text or illustrations in ink from a prepared surface
priorable name
a name that is established and is to be taken into account for purposes of precedence.
priority – See: **precedence**
progeny
the result of propagation.

protologue
   everything published in connection with a name or epithet upon its first publication.

provenance
   the known geographic origin of plants or seed, used mainly by foresters to describe worthwhile selections from indigenous populations.

publication
   a principle of nomenclature in a *Code* whereby certain rules must have been fulfilled before establishment is assessed, usually achieved by the distribution of dated printed matter, so as to make it available to the community: the act or process of distributing printed matter through sale, exchange or gift into the public domain.

publish
   to issue a publication: to make public in such work any names, epithets or other nomenclatural acts.

published
   of a name or epithet, one that fulfills the requirements of publication.

publishing author
   the author who is principally responsible for a contribution published in a book, periodical, etc., and to whom included names or nomenclatural acts are to be attributed.

quotation marks
   marks used to enclose one or more words: double quotation marks ("...") may be used to indicate quoted passages from one text in another, to indicate original spellings or names or epithets that are not established: single quotation marks ('...') are placed around the epithet of a cultivar.

ramet
   an individual member of a clone.

rank
   any category in the nomenclatural hierarchy.

registered name
   a name or epithet that has been entered in a register: one accepted by an International Registration Authority.

registered trade-mark
   a trade-mark which has been formally accepted by a statutory trade-mark authority, distinguished by the international symbol ®.

registration
   the act of recording a new name or epithet with a registration authority.

reject
   to set aside a name or epithet of a taxon in favour of another name or epithet.

rejected name or epithet
   one that is a result of failure to comply with certain provisions of a *Code*: one that is so designated as a result of another being sanctioned by an International Registration Authority.

reprint – See: **separate**

retroactive
> active back in time: nomenclaturally, unless expressly limited, a provision in a *Code* is operational regardless of when it became part of that *Code*, and applies equally to any and all names and epithets proposed after the accepted date at which establishment in that group begins.

rogue – See: **off-type**

roguing
> the removal of living off-types from a crop.

roman
> the printers' font that is not italics.

rootstocks
> the living material upon which a scion is grafted which, at some point in time, is reduced to leave only root tissue beneath the scion.

rules
> regulations (often set out as Articles) in a *Code* which must be followed.

sanctioned epithet
> an otherwise unacceptable cultivar epithet made to be the accepted epithet by action of an International Registration Authority, under certain limited conditions.

scientific name
> the name of a taxon formed and maintained under the rules of the international codes of nomenclature.

scion
> the shoot containing buds that is used for grafting.

seed bank
> a place where seed is held for safe keeping.

selection – See: **maintenance**

self-pollination
> the transference of pollen from the male part of a flower to the female part of a flower on the same plant.

selling name – See: **trade designation**

sensu lato
> used to mean "in a broad sense".

separate
> a part of a periodical or other work, often the contribution of a single author, printed separately from the regular issue of the main work, and usually intended for private distribution by the author: a reprint.

serial
> a publication issued at regular or irregular intervals with no scheduled termination.

series
> a term used in seed-marketing to denote an assemblage of cultivars based on a certain ideotype and differing from each other usually only in one character, normally flower colour.

series mixture
> a term used in seed-marketing to denote an assemblage of a mixture of cultivars from a Series.

sexual propagation
in plants, the propagation of organisms by means of seed following fertilization.

S₁ Hybrid – See: **synthetic hybrid**

siglum (plural: sigla)
a made–up word derived from combining abbreviations of other words.

silviculture
the science of forestry and the cultivation of woodlands for commercial purposes.

simultaneous publication
the publication of two or more names for a single taxon in the same work by the same author.

single cross – See: **F₁ Hybrid**

species (*species*)
the basic category in the nomenclatural hierarchy.

specimen
a plant, or part of a plant, preserved as a unit for scientific study.

specioid – See: **cultigen**

spelling
the choice and arrangement of the letters that form a word.

split
the division of a taxon into two or more taxa.

sport
a mutation which has occurred on part of a plant.

stability
a state in which change is uncommon: in nomenclature, the maintainence of names and epithets which are in use.

Standard
a specimen, seed sample or illustration kept and maintained to demonstrate the diagnostic characteristics of a cultivar.

Standard portfolio
a device in which a Standard and allied information are kept together.

starting point
the date on which establishment of names in a particular group begins.

statutory epithet
an epithet denominated by a statutory plant registration authority.

statutory plant registration authority (SPRA)
an organization established by legal enactment of a particular country or by a legal treaty between countries.

statutory registration
registration by a statutory plant registration authority.

strain
a confused term having several meanings; in cultivated plants, often referring to a seed–raised cryptic variety. See also: **maintenance**

subspecies (*subspecies*)
the category in the nomenclatural hierarchy between species (*species*) and variety (*varietas*).

subspecific epithet
that used following the indication of rank to designate a subspecies.

suffix
a letter or group of letters added to the stem of a word.
superfluous name
an unacceptable name whose name–bearing type is that of an earlier name.
symbol
a mark or sign which is appended to a name, which is in itself not part of that name, yet which provides extra information about the taxon concerned.
synonym
a name or epithet denoting a taxon in a given taxonomic position which, except in certain circumstances, is not the accepted name or epithet.
synonymy
a list of synonyms.
synthetic hybrid
a plant–breeding term for the result of open–pollination within a number of defined, controlled lines.
syntype
under the *ICBN*, any one of two or more specimens cited in the protologue when no holotype was designated, or any one of two or more specimens simultaneously designated as types.
tautonym
the same word repeated more than once in a name.
taxoid – See: **culton**
taxon (plural: taxa)
the international abbreviation for the words "taxonomic group(s)"
Taxon
the official journal of the International Association for Plant Taxonomy.
taxonomic category
a subdivision in a hierarchical system
taxonomic group
the group into which a number of similar individuals may be classified.
teleological classification
one that defines sets according to degrees of usefulness.
teratological plant
a deformed or otherwise grossly abnormal plant.
teratology
the study of abnormalities or monstrosities.
termination
an inflection; the part of a word added to a Greek or Latin stem when the word is inflected.
tetraploid
a polyploid with four sets of chromosomes.
thesis (plural: theses)
a proposition laid down or stated: a dissertation to maintain and prove this.
three–way cross
the repeatable crossing of three F_1 hybrids.
top–cross
a plant–breeding technique in which cultivars are compared by crossing them all with the same pollen parent, and then assessing the progeny. The pollen parent will be a line.

topophysic plants
   two or more plants with the same genotype but differing in their phenotype
   due to being originally asexually propagated from different parts of the
   same parent plant.
topovariant
   a distinguishable group of plants from a given provenance.
trade designation
   an epithet which, although technically superfluous, is used to market a plant
   when the original epithet is considered unsuitable for selling purposes.
trade–mark
   any sign (usually made from words, letters, numbers or other devices such
   as logotypes) that individualizes the goods of a given enterprise and distinguishes them from the goods of its competitors.   See also: **registered
   trade–mark**
translation
   the rendering of words from one language to another.
transliteration
   the rendering of letters or signs from one writing system to another.
transcription
   to copy verbatim from one written work to another: the rendering in written
   form of sounds of human speech, especially of languages which have no
   written form.
transformed plant – See: **genetically modified plant**
trinomen
   a binominal name followed by an epithet at some infraspecific rank.
trinomial – See: **trinomen**
triploid
   a polyploid with three sets of chromosomes.
trivial name
   a specific epithet.
type – See: **nomenclatural type**
typification
   the act of designation or selecting a type for a name.
typographical error
   an error introduced by a printer.
typography
   the visual presentation of printed words and numbers.
undetermined
   of a specimen, not identified.
UNESCO
   the acronym for United Nations Educational, Scientific and Cultural Organization.
UPOV
   the international abbreviation for the 'Union Internationale pour la Protection des Obtentions Végétales' (the 'International Union for the Protection
   of New Varieties of Plants'), the international body charged with administ-
   ering Plant Breeders' Rights.

USLC
the acronym for the United States Library of Congress.
valid publication – See: **establishment**
validly published epithet/name – See: **epithet/name**
variant
a plant or group of plants which shows some measure of difference from the characteristics associated with a particular taxon.
varietal denomination
the name for a taxon designated by a statutory plant registration authority. See also: **statutory epithet**
variety
term used in some national and international legislation to denominate one clearly distinguishable taxon from others; generally, in such legislative texts, a term exactly equivalent to cultivar.
variety (*varietas*)
the secondary category in the nomenclatural hierarchy between species (*species*) and form (*forma*).
vegetative propagation – See: **asexual propagation.**
vernacular name
a translation of a scientific name into a local language. See also: **colloquial name** and **common name**
voucher specimen
a nominated specimen representing the plant or taxon mentioned in a text.
vowel
in a particular language, speech–sound produced by vibrations of vocal cords, modified or characterized by form of vocal cavities, but without audible friction: letter representing this such as, in English, a, e, i, o, u when used on their own and y when used with a consonant. See also: **connecting vowel**
wild plant
one that originated in the wild. See also: **indigen**
witches' broom
a mass of congested, often stunted, stems and foliage on a plant caused by genetic malformation in the growing shoots.
word
vocal sound or combination of sounds, or written or printed symbols of these, constituting a minimal element of speech having a meaning as such, and capable of independent grammatical use. See also: **compound word**
word element
a component part of a word: the parts of a word separated by hyphens.
work
in nomenclature, any pertinent written information, whether published or unpublished.
xerography
any method of producing numerous identical copies based upon an electro-static process such as by photocopying or laser printing.

Index to Scientific Names

Subject Index

The references in this index are not to pages but to the Articles, Recommendations, etc. of the *Code*, as follows: Div. = Division; Pre. = Preamble; Prin. = Principles; arabic numerals = Articles or, when followed by a letter, Recommendations; Ex. = Examples; N. = Notes; fn = footnotes; App. = Appendices.